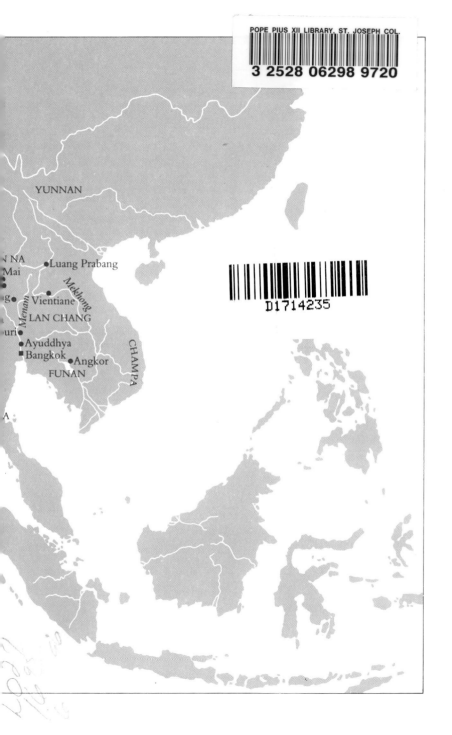

IMAGES OF ASIA

Voyage of the Emerald Buddha

Titles in the series

Balinese Paintings
A. A. M. DJELANTIK

Bamboo and Rattan
JACQUELINE M. PIPER

Betel Chewing Traditions
in South-East Asia
DAWN F. ROONEY

The Birds of Java and Bali
DEREK HOLMES and STEPHEN NASH

The Birds of Singapore
CLIVE BRIFFETT and
SUTARI BIN SUPARI

The Birds of Sumatra and Kalimantan
DEREK HOLMES and STEPHEN NASH

Borobudur
JACQUES DUMARÇAY

Burmese Puppets
NOEL F. SINGER

Early Maps of South-East Asia
R. T. FELL

Folk Pottery in South-East Asia
DAWN F. ROONEY

Fruits of South-East Asia
JACQUELINE M. PIPER

A Garden of Eden: Plant Life in
South-East Asia
WENDY VEEVERS-CARTER

Gardens and Parks of Singapore
VÉRONIQUE SANSON

The House in South-East Asia
JACQUES DUMARÇAY

Images of the Buddha in Thailand
DOROTHY H. FICKLE

Imperial Belvederes: The Hill Stations
of Malaya
S. ROBERT AIKEN

Indonesian Batik
SYLVIA FRASER-LU

Javanese Gamelan
JENNIFER LINDSAY

Javanese Shadow Puppets
WARD KEELER

The Kris
EDWARD FREY

Life in the Javanese Kraton
AART VAN BEEK

Mammals of South-East Asia
EARL OF CRANBROOK

Musical Instruments of South-East Asia
ERIC TAYLOR

Old Bangkok
MICHAEL SMITHIES

Old Jakarta
MAYA JAYAPAL

Old Kuala Lumpur
J. M. GULLICK

Old Malacca
SARNIA HAYES HOYT

Old Manila
RAMÓN MA. ZARAGOZA

Old Penang
SARNIA HAYES HOYT

Old Singapore
MAYA JAYAPAL

Rice in South-East Asia
JACQUELINE M. PIPER

Sarawak Crafts
HEIDI MUNAN

Silverware of South-East Asia
SYLVIA FRASER-LU

Songbirds in Singapore
LESLEY LAYTON

Spices
JOANNA HALL BRIERLEY

Voyage of the Emerald Buddha
KAREN SCHUR NARULA

Voyage of the Emerald Buddha

KAREN SCHUR NARULA

KUALA LUMPUR
OXFORD UNIVERSITY PRESS
OXFORD SINGAPORE NEW YORK
1994

Oxford University Press

Oxford New York Toronto
Delhi Bombay Calcutta Madras Karachi
Kuala Lumpur Singapore Hong Kong Tokyo
Nairobi Dar es Salaam Cape Town
Melbourne Auckland Madrid

and associated companies in
Berlin Ibadan

Oxford is a trade mark of Oxford University Press

Published in the United States
by Oxford University Press, New York

© Oxford University Press 1994
First published 1994

All rights reserved. No part of this publication may be reproduced,
stored in a retrieval system, or transmitted, in any form or by any means,
without the prior permission in writing of Oxford University Press.
Within Malaysia, exceptions are allowed in respect of any fair dealing for the
purpose of research or private study, or criticism or review, as permitted
under the Copyright Act currently in force. Enquiries concerning
reproduction outside these terms and in other countries should be
sent to Oxford University Press at the address below

British Library Cataloguing in Publication Data
Data available

Library of Congress Cataloging in Publication Data
Narula, Karen Schur.
Voyage of the Emerald Buddha/Karen Schur Narula.
p. cm. — (Images of Asia.)
Includes bibliographical references.
ISBN 967 65 3057 3:
1. Emerald Buddha (Statue) 2. Gautama Buddha—Art.
3. Sculpture, Buddhist—Thailand. 4. Sculpture, Thai.
I. Title. II. Series.
NB 1912.G38N37 1994
739'.9593—dc20
94–4619
CIP

Typeset by Typeset Gallery Sdn. Bhd., Malaysia
Printed by Kim Hup Lee Printing Co. Pte. Ltd., Singapore
Published by Oxford University Press,
19–25, Jalan Kuchai Lama, 58200 Kuala Lumpur, Malaysia

*For my parents and Arruvind,
and Nikhil-Alexei,
in love and appreciation*

Preface

EACH day hundreds of the faithful, the awed, and the curious from all corners of the world pause within the magnificent chapel in which, since 1784, is housed Thailand's most famous and revered image of the Buddha. Kings have coveted it, firm in the belief that possession would bring the blessings of peace and prosperity upon the inhabitants of their realms. For more than two hundred years, the Emerald Buddha has kept watch from its exalted position in the Thai capital. In return, it receives supreme reverence and the accolade of 'palladium of the nation'.

For the curious, questions arise. Where did it come from? Who sculpted the precious image? Speculation is rife. Its origin has been attributed to ancient India, Sri Lanka, and Thailand itself. However, none of these possible birthplaces can be confirmed with irrefutable certainty.

This book takes the liberty of sailing through waters wherein fact and legend are blended in further speculation. From its roots in the teachings of a man who lived two and a half millenniums ago, Buddhism has grown, spreading its branches to lands distant from its original home in India. Today, the island of Sri Lanka and four of the countries of South-East Asia are strongholds of the Theravada Buddhist faith. This book offers a peek at the panorama of those countries' histories through the speculative voyage of the Emerald Buddha. And even if no definite conclusions can be drawn from the journey, I hope it may serve to illustrate some of the strengths, frailties, and human endeavours of the world within which the Emerald Buddha has made its home.

Bangkok KAREN SCHUR NARULA
June 1993

Contents

	Preface	*vii*
1	Introduction	*1*
2	India	*9*
3	Sri Lanka	*25*
4	Burma	*36*
5	Cambodia	*41*
6	Conundrum	*56*
7	Thailand	*59*
8	Laos	*70*
9	To Bangkok	*77*
10	Epilogue	*85*
	Select Bibliography	*87*

1
Introduction

Being a description of the changing of the clothes ceremony as performed by His Majesty the King of Thailand upon the sacred image of the Emerald Buddha, as well as an introduction to the mysterious Chronicle of the Emerald Buddha.

THE great Visechaisri Gate stands with portals flung wide open, stately entrance through a thick wall of white surrounding the two-century-old grounds of the Grand Palace. Under its spired roof pass both Thai and foreign visitors to the monument-studded complex. Today, the first day of Buddhist Lent, the destination of most is the eastern section of the compound, the Chapel Royal, or as it is commonly called, the Temple of the Emerald Buddha. In but a short space of time, His Majesty the King of Thailand will arrive to carry on the tradition of great kings before him.

In galleries on three sides of the *ubosoth*, the ordination hall where the Emerald Buddha is enshrined, are the fine murals of the *Ramakien*, the Thai version of the Hindu epic, the *Ramayana*. On the northern side are the Phra Mondop (Library) and Royal Pantheon. Now, on the stone floor of the courtyards between these several structures, the crowds are building up. Many sit on straw mats, feet pointing delicately away from the sacred site. Parasols, of hues nearly as varied as the glazed Chinese tiles set into the low boundary walls surrounding the *ubosoth*, are raised to protect from either the sun or rain. For today, on the full moon in July, the Emerald Buddha will be clad in the garments of the rainy season and the skies are yet undecided.

The air is sweet with the scent of incense. Lying between two small *sala*, or pavilions, on the eastern side of the *ubosoth*, is an enclosure wherein rest images of the Buddha as well as two statues of cows, vestiges of Brahman influence on the Buddhism of

1. Mural of Wat Phra Keo in Wat Benchamabopit. (Luca Tettoni/Photobank)

Thailand. Dark vessels of oil harbour dancing flames. Worshippers gather up candles and sticks of joss, dip them into the fire and, as the smoke snakes upwards, bow over them before an image of the Buddha. Softly, like bird wings brushing as flocks take flight, the whisper of their prayers is carried on the wind. Along with stalks of flowers, the fragrant offerings are then placed into great pots.

More and more people fill the area. Black-bereted soldiers thread through the mass, gently monitoring the growing current of excitement. Spurred by the gathering wind, tiny bells with heart-shaped flappers tinkle from the roofs. A group of monks passes by, clad in saffron robes whose style of draping is barely different from that seen on second-century Buddha images from Gandhara in the north-west of the Indian subcontinent. On the golden tiers of the giant stupa, Phra Si Ratana Chedi, rising to the west of the Phra Mondop, sit rows of pigeons with feathers as grey as the thickening clouds. Babies, growing restless within their mothers' arms, are suddenly captivated by the arrival of the royal musicians into one of the *sala* to the east.

The wind swells. The crowds sense the approach of His Majesty. The charge of excitement at being in his presence ripples across the sea of bodies. The white uniforms of the officials part it, making way for the King as the bells ring and the pigeons lift off from the *chedi*. Royal umbrellas shield His Majesty as he bends to receive offerings of flowers from his subjects. And when the King ascends the steps of the *ubosoth*, the darkened skies burst and at last the rain begins to fall.

High within the ornate interior of the *ubosoth*, seated with its right leg resting on the left in the hero pose of *virasana* atop a gold-covered wooden throne, rests Phra Keo Morakot, the Emerald Buddha. The golden altar, or *busabok*, upon which the throne sits represents the aerial chariot Pushpaka of the Hindu gods. His Majesty mounts steps at the rear, reappearing at the level of the Emerald Buddha. He removes the Hot Season garments. With a cloth, following the traditions of long-gone kings, His Majesty now begins carefully, painstakingly, to wipe the smooth jade of the image. At this moment, from within the pavilion outside, the music explodes in a jubilation of xylophones,

2. His Majesty King Bhumibol Adulyadej (Rama IX) and Crown Prince Maha Vajiralongkorn at Wat Phra Keo. (Luca Tettoni/Photobank)

horn, and drums. Like the summons of a giant conch shell, the horn crescendos intermittently. Once polished, the Emerald Buddha is crowned with a headpiece of a golden *ushnisha*—one of the thirty-two marks characterizing the Buddha's extraordinary nature as described in Buddhist scriptures—with gentle curls beset with sapphires, topped with an enamelled flame finial. There follows a robe of gold decorated with rubies. His Majesty smoothes it down, ensuring that it sits well, keeping one hand upon the image's shoulder while, with the other, adjusting the wrap.

Down below, on the geometrically patterned marble floor of the *ubosoth* where the officials of state and religion wait, candles are ritually passed in a circle, their flames visible to the spectators outside.

The rain has stopped; blue is returning to the sky. The guards allow some of those waiting to move into the enclosure between

3. The Emerald Buddha, clad in the garments of the Hot Season. (Luca Tettoni/Photobank)

the *ubosoth* and the low boundary walls. The music ends abruptly and His Majesty descends from the hall. In former times, only princes and officials were so blessed, but today the King of Thailand sprinkles lustral water upon all of his subjects who have found places within the enclosure. Their humble faces reflect the joy of their belief in the benevolence of HM King Bhumibol Adulyadej, King Rama IX, and in the enduring sanctity of the Emerald Buddha.

Chronicle of the Emerald Buddha

As yet, historical documentation of the Emerald Buddha reaches back only to northern Thailand's Chiang Rai in the year AD 1434 when a Buddha statue covered with stucco was found there inside a *chedi* which had been struck by lightning. The image was subsequently transferred to the residence of the temple's abbot who one day discovered that some of the stucco had flaked off. Underneath was hidden a green image. The stucco was duly removed and thus the Emerald Buddha was revealed.

That the Emerald Buddha had long been the object of intense veneration and interest can be attested to by the existence of several manuscripts purporting to trace its origins. These have been found not only in Thailand but also in Laos and Cambodia. All of them seem to have been based on an original text which has been lost. And while there are in the various manuscripts omissions and differences, enough similarities remain to realize that the original version must have been considered of great importance.

The Chronicle of the Emerald Buddha as recorded on palm leaves in one of the manuscripts discovered, that of Chiang Mai, tells a tale in which are fused historical facts and religious legend, a tale in which the tapestry of time is woven in patterns so overlapping that the weave is often a tight mass of inseparable threads. Chronicles are, after all, the writing down of events and legends which were at first preserved only orally. When at last the accounts were recorded, usually too much time had passed. Events had been forgotten or distorted. Fact and fiction fuse in confounding ways, so that caution must be exercised in accepting chronicles as historical evidence. Yet, despite the contradictions and chronological impossibilities, and despite the rather unscholarly piecing together of fantasy and reality, there remains a certain inviolable truth. The Chronicle embodies a belief. And it is belief, as faith, which forms the backbone of religion. To even attempt to unravel the mystery of the Chronicle, and of the revered Emerald Buddha, it is necessary to go beyond the mere words. One must travel an airy path through the natural dimensions of religion and myth, history and geography, to behold a picture that is greater than the sum of its parts.

The Chronicle opens with the *parinirvana*, or physical death, of Siddhartha Gautama, or Shakyamuni, the man who became known as the Buddha and upon whose life and teachings is based one of the world's great religions. The *parinirvana* of the Buddha took place when he was eighty, somewhere between the years 563 and 483 BC (in Theravada tradition, it occurred in 543 BC). According to the Chronicle, it was in 44 BC, in the ancient imperial capital of Pataliputra in India, that the events leading to the creation of the Emerald Buddha took place. That at least two of the figures first mentioned, King Milinda (the Bactrian King Menander) and the Buddhist sage Nagasena, had lived some hundred years before, need not, as previously stated, destroy the fabric of the Chronicle. Thus, to begin, the Chronicle's first Epoch relates how Nagasena conceived the idea of creating an image of the Buddha to encourage the flourishing of Buddhism.

He was aided by the Aryan god of thunder and battle, Lord Indra, who, together with the celestial architect Visukamma, went to procure a certain precious stone from under the guard of a thousand genii on Mount Vipulla. The genii were reluctant to part with that particular stone as it belonged to the Universal Monarch. They proposed in its place a magnificent green gem, 'four times the size of the fist and three fingers in width, and almost one cubit one hand in length'. Once in possession of the jewel, Nagasena wondered who would carve the image. Visukamma came to his aid in the guise of a sculptor.

Taking the precious stone away, he went to the realm of angels where celestial beings carved it in seven days and seven nights. Under Indra's orders, Visukamma returned to earth, carrying the image upon his shoulders; Nagasena was overjoyed by the figure of the Buddha.

On the night of its consecration, the moon was full. Under its rays, the statue was given the title of Phra Keo Amarkata Sing Dieo, or the Unique Emerald Buddha. Nagasena then brought out a golden vessel containing seven relics of the Buddha and made an invocation: 'May these relics go into this Buddha if for the blessing of all men and angels, his image is to last as long as five thousand years.' The relics then entered the statue, and Nagasena continued: 'This image of the Buddha is assuredly going to give to religion the most brilliant importance in five lands. . . .'

4. Lord Indra. (Drawing by Thaworn)

Buddhism did indeed spread to countries beyond the borders of India, and Theravada Buddhism, in particular, to Sri Lanka, Burma, Cambodia, Laos, and Thailand. Just what part the Emerald Buddha played in its propagation is a challenging mystery.

2
India

Being an observation of the early years of Buddhism in India and the rise of Buddhism, as well as mention of a few persons, events, and trends of importance.

AT the time of the Emerald Buddha's legendary creation, the lands now known as India had for centuries already felt the footsteps of diverse peoples. The earliest traces of these nameless souls of the Stone Age date from between 400,000 and 200,000 BC. Thousands of years later, many had already learned how to live together in villages, taming both the earth and some of its animals. Small cultures flourished and then faded with time, leaving behind bricks, relics, and the genetic patterns for those cultures which would follow. The societies and beliefs which evolved from the earliest did so gradually, ever adapting and modifying, creating new systems and concepts as mankind's own understanding of the universe developed.

Our curtain on the historical stage opens during the third millennium BC. Primitive man had matured, and civilization as the world had never known sprang into being in the valleys of three great rivers. Yet, unlike the wealth of written material which the nearly contemporary great civilizations of the Nile and Euphrates River valleys left behind, only scanty inscriptions serve to highlight the people of the Indus, or Harappa Culture.

While other settlements on the subcontinent continued in transitional stages of development, the Indus cities, of which Harappa and Mohenjo Daro were the most important, attained a level of civic planning which was astounding in its sophistication. Excavations suggest that the cities were governed by a centralized state. The thriving economy was probably due to surplus agricultural produce and the income derived from trade. This existed both with other areas of the subcontinent and, from the evidence of stamp seals and other items found in ancient

5. Indus figures and seals. (National Museum, New Delhi)

Ur, with the inhabitants of Mesopotamia and the Persian Gulf.

The religion of the Harappa people embodied several elements, some of which would resurface centuries after in later Hinduism. The horned deity, the Mother Goddess, the bull, phallic worship, and certain trees—notably the pipal (bodhi)—were evidently all particularly sacred. Had the Harappa Culture continued in its organized, comfortable way, its mysterious script might have evolved into but another of contemporary India's many languages, allowing historians to better document one of antiquity's most astonishing civilizations. But by about 1600 BC, at the fall of the First Babylonian Dynasty, the Harappa Culture, too, was on its way to disintegration.

Several elements contributed to the disappearance of this great civilization, among them floods, the drying up of ancient river beds, and invasions by barbarians. While these latter intrusions had been taking place sporadically over the centuries, it was the migratory movements of the Aryans which ultimately ended the Harappa Culture. These tribes were descendants of an earlier people from eastern Europe, the Indo-Europeans, who had, over the centuries, migrated to the western reaches of Europe as well as to the steppes of Central Asia. Only now rightfully bearing the name of Aryan, from here the tribes further divided, to thunder into India by horse and chariot from Iran.

For the first several centuries of their expansion in India, cattle-rearing remained the chief occupation of the Aryans. The cow itself was the unit of currency and, thus, while necessarily limiting the distances travelled for trading, was often the object of intertribal wars due to cattle-stealing. The occasional consumption of beef was permitted by Aryan society. It is possible to speculate that the economic value of the cow was a basis for its subsequent sanctity in later Hinduism.

Over time, the tribes moved eastwards from the Indus and the Punjab, along the foothills of the Himalayas, and into the valley of the Ganges, clearing forests and colonizing, founding agricultural settlements. The south would only later come to feel the effects. The discovery of iron in about 700 BC accelerated the changes in lifestyle from semi-nomadic pastoralism to village communities. And during the last stage of Aryan expansion, urban life itself was reborn, albeit in a small way.

The Aryans had arrived in India equipped with a rudimentary social hierarchy in their tribal structure. This would ultimately metamorphose into an involved class system, based initially on colour. Perhaps fearing the loss of Aryan identity through assimilation with the indigenous peoples (the Dasas), the Aryan

6. Detail of Krishna Mandapa, Mahabalipuram. (Bangkok National Museum)

priests, or Brahmans, evolved a system of dividing society into three classifications based on profession, and one on a mixture of race as well as occupation. Over the centuries this would ultimately evolve into a complex system of sub-castes.

The role of the priest in Aryan society was vital. It is from the wealth of words initially memorized and later penned by these powerful guardians of Vedic lore that India's history first comes to life, illuminating a treasure trove of philosophical richness. Hinduism would owe much of its origin to the social institutions and religion of the Vedic culture and Buddhism would owe much to the refutation of these same keepers. The Vedas were followed by the Sutras, Brahmanas, and later the Upanishads. Collectively, these hymns and treatises were manuals of worship and social laws. From a primitive animism filled with deities of which some had their roots in the Aryans' Indo-European past, to the rituals of sacrifice from which the priests and to a lesser extent the kings gained increasing power, the Aryans' conception of the universe gradually developed.

Hymn of Creation
Then even nothingness was not, nor existence.
There was no air then, nor the heavens beyond it.
Who covered it? Where was it? In whose keeping?
Was there then cosmic water, in depths unfathomed? . . .
But, after all, who knows, and who can say,
Whence it all came, and how creation happened?
The gods themselves are later than creation,
So who knows truly whence it has arisen?

Rigveda

While initially death brought either heaven or hell, later hymns suggest the rebirth of souls in plants. By the seventh and sixth centuries BC, a new concept was gaining ground. This was the transmigration of souls, in which those who had lived lives of good conduct would be reborn as humans while the unrighteous would suffer reincarnation into what were considered lowly animals. Not only humans were singled out for transmigration, however. The gods themselves were said to pass away, their places taken by other gods.

This theory of just rewards would evolve into the doctrine of karma, wherein the deeds of one lifetime were thought to effect the quality of existence in the next. The transmigration of souls thus was seen as the explanation behind human sufferings. While such a system was undoubtedly a comfort to many, there were those who remained dissatisfied. For them, there needed to be an exit from the endless cycle of birth and rebirth. Many answers were found in asceticism and mysticism.

Several factors contributed to making the seventh and sixth centuries a time of vigorous intellectual inquiry and speculation. Aryan society had changed, the old, solid structure of tribal communities fragmenting with the formation of republics and power-seeking monarchies; the amount of communally owned land was on the decline. At the same time, trade with other areas was intensifying. A great new social order was coming into being. Ascetics strove in an astounding variety of ways to interpret the mysterious workings of the cosmos.

Yet, not all those who quested for knowledge isolated themselves from their communities. Many returned to ask questions, and propound their own newly conceived theories. The challenge to the existing religious and social norms was enough to cause the Brahmans concern. Ever quick to react to a situation by minimizing threats or maximizing benefits to themselves, the priests accepted some of the theories, incorporating them into orthodox belief. Thus were the Upanishads later to become accepted and asceticism itself was officially pronounced one of life's four stages.

However, not all sects and doctrines could be harmonized with the orthodox. It was during this time of metaphysical exploration that one of the world's most influential thinkers, Siddhartha Gautama, or Shakyamuni, attained Enlightenment under a bodhi tree in Bodhgaya. Like his fellow Hindu reformer Mahavira, founder of Jainism, the Buddha did not seek to shatter the framework of contemporary Hindu society. In what would have future influence on the Buddhism of Thailand and other countries of Asia, followers of the Buddha were not asked to abandon their belief in Hindu deities or familiar superstitions.

Nevertheless, the Buddha rejected the authority of the Vedas and the practice of animal sacrifices. Maintaining that a man's position in society was determined not by birth but by conduct and character, caste and the hereditary superiority of the Brahmans were dismissed. Anyone could follow the Path. This would be facilitated in centuries to come, through a command of the Buddha that sermons be delivered not in Vedic verse but in the vernacular. Thus, when the monks went forth to spread his teachings, and when at last these were written down some centuries later, the people received the message in the languages of their own lands.

Some of the Buddha's premises were concepts which had already evolved over the preceding centuries. The strength of the Buddha's message lay in the breadth of its appeal. Even Brahmans

7. Bodhgaya stupa. (Luca Tettoni/Photobank)

would soon be numbered among Buddhism's adherents. The Buddha revealed his conceptual formulations of the nature of existence in the Dharma, the term which encompasses, amongst other aspects, the Law, Truth, and Doctrine of Buddhism. Through a system of ethical thinking and living, including non-violence and the avoidance of extremes, the spiritual journey to Nirvana, freedom from the wheel of rebirth, was available to all. The existence or non-existence of a God was not regarded as essential to the universe or the system.

In the centuries after the Buddha's death, what had begun locally as a novel sect of Hinduism would gradually develop into a dynamic system of complex beliefs going further and further afield. And while Hinduism would take and make into its own ever evolving entity something of Buddhism's spirit, Buddhism itself would evolve into different sects, each stressing the importance of its own distinctive tenets as transformed by the interpretations or needs of regional Orders of Monks.

While at this time much of India was still populated by forest tribes, civilized settlements along the Ganges and Indus and their plains, as well as in the south, were growing in number and size, although not without rivalry. Ultimately, the Kingdom of Magadha, two of whose kings, Bimbisara and Ajatasatru, had been contemporaries of the Buddha, gained ascendancy, establishing itself as the centre of political activity in the north-east. During the ensuing century, Pataliputra, the first city mentioned in the Chronicle of the Emerald Buddha, would become its capital, a city of grandeur.

In 326 BC, the crossroads city of Taxila, then still a centre of Vedic learning in the north-west, welcomed a man whose short visit to India would have appreciable effects on the evolution of Buddhist art. Although the invasion of Alexander of Macedonia was not deemed important enough in itself to be mentioned in any Indian texts, this earliest Hellenistic acquaintance would open the doors for subsequent visits and influence from the Western world in the centuries to come.

Contemporary with Alexander was Chandragupta Maurya, usurper of the Pataliputra throne and historically India's first emperor. Both he and his son Bindusara, advised by their Brahman minister Kautilya, were to spend their reigns conquering territories from coast to coast and far into the south. Diplomatic contacts existed between the Mauryan capital and Seleukos Nikator's Syria, as well as with Persia and Ptolmeic Egypt. But it was trade which was the most important link binding communications, both foreign and internal, and luxury imports from the south, including today's Sri Lanka, were especially coveted. The Mauryan government exercised strict supervision over private trade and indeed its citizenry as a whole.

While much of the subcontinent had come under Mauryan suzerainty, one strategic area on the east coast remained defiant. Around 261 BC, deploying the military superiority of his forefathers, the Emperor Ashoka conquered Kalinga in a brutal campaign. It was this event which brought Buddhism from out of the shadows and into the spotlight of an empire. Had a lesser man experienced his sorrow, it is doubtful that the results would have been comparable. But the Emperor's remorse for the sufferings of the Kalingan people gradually brought about his devotion to Buddhism, and through his subsequent support, the erstwhile sect began its ascent into a world religion.

Seeking to actualize his own affirmation of the Buddha's saying that all men were his children, Ashoka during his reign augmented the social services of his empire. By setting an example of enlightened government and the discarding of aggressive warfare, he hoped to convince the world of the righteousness of Buddhist ethics. Ashoka's Dharma could be seen as an attitude of social responsibility in honour of the dignity of man. Thus, despite the Emperor's personal devotion to his particular interpetation of Buddhism, he advocated tolerance of other creeds. In Ashoka's capital, too, animal sacrifices were banned, and vegetarianism began to gain in popularity.

The Emperor, whose dominions covered the subcontinent from as far as the Hindu Kush in the north to the border with the independent Tamil states in the extreme south, sought to educate his subjects in a remarkable fashion. From the medium, positioned in scattered locations across the empire, it is apparent

8. Detail of an Ashokan sandstone pillar, Sarnath. (Bangkok National Museum)

that a sufficiently high percentage of the population was literate. Carefully crafted messages and moral codes extolling a virtuous life were inscribed on both rocks and stylized pillars, the latter at last illuminating in stone, rather than perishable wood, the expert artistry of the Indian sculptor of the age.

While trade and diplomacy continued as motivators of foreign intercourse, Ashoka's zeal for the propagation of Buddhism meant that religious missions were sent to distant lands. Those to the Hellenized kingdoms of western Asia and Africa met with ultimate failure; the Emperor's greatest contemporary success lay to the south, on the island of Sri Lanka. Borne on the winds of Ashoka's convictions and the efforts of missionaries over centuries to come, Buddhism would spread to other countries as well; stupa and *vihara*, or monasteries, would come to dot diverse landscapes. But in its journey to distant lands, the nature of Buddhism would of necessity change. The Indian essence of its origins was transformed as the religion adapted itself to the intrinsic needs of a wide variety of peoples.

Ashoka's heirs were ousted by the Hindu King Pushyamitra around 185 BC, after which Buddhism was for several centuries no longer royally favoured. Dynasties of varying durations and religious persuasions followed, the centres of power shifting in their wake. The finest artistic legacies of this time are, however, Buddhist creations. Although Ajanta would know full flowering in the Gupta Age, the rock-cut Buddhist complexes of *vihara* and *chaitya*, or places of worship—sanctuaries situated in the Western Ghats—were the unique creation of the second century BC faithful. It would take longer for the architects and sculptors of the far south to leave their wooden medium for the sound of hammer and chisel against rock; this would occur some seven centuries later during Pallava rule.

During Ashoka's reign, two Seleukidan provinces had declared independence as Parthia and Bactria. The latter is of interest to us here as it was the Bactrian King Menander, whose capital was in the Punjab from about 160 to 140 BC, who is mentioned in the Chronicle as the Milinda who was converted to Buddhism by the philosopher-monk Nagasena. This was later recorded in a Pali text called *Questions of Milinda*. It is this same Nagasena

9. Bactrian coin depicting King Menander. (National Museum, New Delhi)

whom the Chronicle credits with the idea of creating an enduring image of the Buddha: 'The Buddha, the Doctrine and the Church, each represents a gem [the Three Jewels of Buddhism], so I have to get a precious stone with a very great power in it to make the statue of the Lord.' As stated earlier, however, the date in the Chronicle attributed to its creation, 44 BC, leaves a time discrepancy of about a century.

None the less, at the time of Nagasena, influence from the West was beginning to be felt in a way which it had not been during the time of Alexander. The rise of the Graeco-Bactrian kingdoms brought with them some Western theories of science, and perhaps even aspects of the drama of the Greek stage. New perspectives on life may have been shared by both the foreign dynasties and those exposed to the descendants of the Indo-Greeks. Nagasena may indeed have been instrumental in converting King Menander. At the same time, what points of view did the King, in turn, present to his teacher? It does not seem difficult to imagine a philosopher-monk of such powerful and dedicated persuasion being inspired to search for a means of widening and perpetuating adherence to the Buddhist system of belief. And even if Nagasena did not commission a more earthly, unnamed sculptor to create from a piece of probably imported nephrite jade the first likeness of the Buddha, perhaps the seeds of such an idea were sown during his age, to flower in the near future when men would dare to give form to the man behind the message.

The concept of a sacred object possessing the power to preserve the safety of a city or state (although in this case a system of belief) would not have been unknown. The ancient Greeks had Pallas Athena, the statue that they believed assured the safety of Troy as long as it remained within the city. Seen from this perspective alone, the fact that future kings were to seek to harbour a particular image in their realm is understandable. The Emerald Buddha would be such a palladium.

By the time of King Kanishka in AD 120, the focus was on Gandhara in the north-west. Once again, Indian soil had been invaded from the north as the Scythians, pushed out of their homelands by the Yueh-chi migrations from China, sought power first in Bactria, Parthia, and then as far into the interior as Mathura. The Yueh-chi eventually settled into Bactrian lands, the Kushan section of the horde attaining prominence. Power struggles with China occupied the first Kushan kings, the results being the opening up of travel routes through Khotan and along the Taklamakan Desert. King Kanishka was the most noteworthy of the Kushans, a ruler who, like the Emperor Ashoka, gained fame as a patron of Buddhism.

One of the most enduring features of Kanishka's age was the emergence and acceptance of depiction of the image of the Buddha. In the art of Nagasena's time, the Buddha's presence had been conservatively indicated only by symbols, such as a pair of footprints, a vacant throne, or a flaming pillar. A noteworthy remaining example of this is in the carvings adorning the second-century BC stupa at Sanchi. But by the second century AD, the figurative representation of the Buddha

10. Worshipping Buddha's footprints, Madras Museum. (Luca Tettoni/Photobank)

had appeared. It is likely that it developed in both Gandhara and Mathura. Now, at the hands of visionary sculptors, the image of the man became accepted as the icon of the religion. Even Kanishkan coins carried the image of the Buddha. The north-western school of Gandhara itself evolved from those Hellenistic influences as expressed in the philosophy and art of the eastward reaches of the Roman Empire. At the same time, the sculptors of Gandhara contributed to the development of the image of the Bodhisattva, or Buddha-to-be. The Gandhara style would travel the missionary route through Central Asia to China and Japan where it would continue to be used in Buddhist sculpture long after its popularity in India had waned.

Other towns boasted schools of Buddhist sculpture. Amaravati, in the south-east of the Andhra Kingdom of the central Deccan, developed a distinctive and graceful style. Indeed, the earliest examples of Indian art found in South-East Asia are Buddha

11. Amaravati bas relief, Madras Museum. (Luca Tettoni/Photobank)

images of the Amaravati school. A mid-second century AD relief from the Great Stupa there still noted the Buddha's presence through the empty throne and footprints. But by the end of the second century, the Buddha was already in human form. From here on, the products of this intensely vital school would influence the local styles of both Sri Lanka and the South-East Asian countries. Considering its style, South India has been suggested as one of the possible origins of the Emerald Buddha itself.

Being favoured by elements of the rich and powerful, the Buddhist Order had become respected in India. Monkhood no longer meant a life of few material comforts as it had in the years following the Buddha's death. Yet, it was this very support from the affluent which would be one of the factors leading to the eventual decline in the religion's popularity in India. For as the monasteries thrived on their endowments, often secluding themselves in sites away from the general population, this isolation diminished the presence and therefore strength of the religion.

Buddhism itself had been changing over the centuries, for returning with the missions to foreign countries were fresh ways of looking at the teachings. New interpretations of the Buddha's doctrine led to major differences of opinion. At the Fourth Buddhist Council, hosted by King Kanishka in the second century AD, the centuries-old schism between the Hinayana and Mahayana doctrines, the two major sects of Buddhism, was officially recognized.

The demise of the Kushan Empire during the first decades of the third century brought with it the formation of many independent authorities in the region. Pataliputra was taken by the Lichchavis, and the fourth-century marriage of a Pataliputra princess and a raja of Magadha allowed the latter to establish what would become the Gupta Empire. The Guptas apparently appreciated the finer aspects of life, for during their days of power, the arts flourished. Kalidasa penned the play *Shakuntala*, and the carved caves at Ajanta were painted with scenes from the *Jataka* tales, the stories of the Buddha's lives before Buddhahood.

Buddhist sculpture, too, was refined and redefined. Images of the Buddha, evolving from that of suspended action in the Mathura school of the Kushan style, were now crafted with a mood of inner tranquillity.

The Gupta Age has often been said to have produced some of India's finest Buddhist works. At the same time, however, the Hindu art of India was growing more prominent. One response to this was the beginning of esoteric or Vajrayana Buddhism, an outgrowth of Tantric thought which was known to Brahmanism as well. Indeed, many Brahmanical practices themselves became a part of Buddhism. But Buddhism was losing its momentum in India. As it diminished in its homeland, so it gained importance in distant lands as traders as well as missionaries carried the message across mountains and monsoon seas.

12. Gupta-style Buddha, Mathura Museum. (Luca Tettoni/ Photobank)

3
Sri Lanka

Being a perusal of the Second Epoch of the Chronicle of the Emerald Buddha, followed by a brief account of Buddhism's arrival on Sri Lanka, as well as a cursory glimpse at some of the landmarks and people which influenced the island up till the eleventh century.

THE Second Epoch of the Chronicle of the Emerald Buddha opens, its unknown author stating that a civil war had taken place in Pataliputra. Concerned about the safety of the Emerald Buddha, the current ruler felt it wise to send the statue to his friend, the King of Lankadvipa. The plan was that once the fighting had ceased, he himself would journey to collect it. Thus did the Emerald Buddha, accompanied by privy councillors, voyage southwards to be welcomed by the King, who placed it in the Meghagiri. This is presumably the temple at Anuradhapura wherein rain-making ceremonies are said to have been performed.

The King of Pataliputra is likely never to have journeyed to Lanka and thus the statue is said to have remained there for two centuries, bringing fulfilment to the inhabitants and the flourishing of Buddhism on the island. However, forces were at work elsewhere. In AD 457, far away across the seas to the north in Pagan, King Anuruddha was approached by a priest named Silakhandha who voiced his concern about the authenticity of the *Tripitaka* (the three collections of the Buddha's words, Vinaya, Sutra, and Abhidharma), and Buddhist Commentaries used in that city. A request for a fact-finding mission to Lanka, where the Pali commentaries of the legendary Mon monk Buddhagosa were esteemed, was royally sanctioned and Silakhandha and his party of eight priests and two officers set out in two boats. King Anuruddha was possessed of a supernatural ability to fly, and thus the sea voyagers and the Pagan ruler were able to reach the Lankan shores at the same time.

As touched upon in the previous chapter, the time between the decline of the Kushan Empire and the rise of the Guptas saw northern India in a disturbed state. Pataliputra itself came under the control of the Lichchavis, an ancient tribe to which the Buddha himself was related. It may thus not be surprising that Sri Lanka, by then a stronghold of Buddhism, might have been perceived as a haven. The Emerald Buddha would have been considered safe there, protected by people of the faith.

In *Questions of Milinda*, composed as it probably was in the first century AD when trade with the Romans was centred on Eastern luxury goods in exchange for gold, it is said that Indian merchants sailed their vessels as far west as Egypt. They had already been plying the waters to the golden lands of South-East Asia, the ancient Suvarnabhumi and Suvarnadvipa of perhaps Burma, the Malay Peninsula, Indonesia, and Indo-China. The south—and its treasures—was also familiar to them. Setting sail from the Ganges River port of Champa, and later Amaravati, crafts from East India kept the coastline in sight as they rode towards the island to which Sita had been spirited off by Ravanna in the Hindu epic, the *Ramayana*. So too might the Emerald Buddha have set out on its voyage, sheltered inside a vessel which, timbers lashed together for resiliency, creaked and swayed on its way southward along the waves.

Contact between the subcontinent and the already inhabited island had begun some thousand or more years earlier. While legend and historical fact are as often elsewhere fused, it seems plausible that the original home of the first Aryan immigrants to Sri Lanka was the Indus and north-west regions and later the east of India. According to the *Mahavamsa*, Sri Lanka's chronicle of its early history, the arrival from the Vanga country of Prince Vijaya, the founding father, coincided with the death of the Buddha (who was himself said to have visited the island three times). Upon his deathbed, the Buddha was said to have sought celestial protection for the island and its faith. The belief in this would would form the basis of Sri Lanka's concept of itself as 'a place of special sanctity for the Buddhist religion'.

While the Indo-Aryans had brought with them some form

of Brahmanism, and the influences of Hinduism would long be felt, the early teachings of the Buddha eventually found their way across the water as well. But it was not until the days of Ashoka and his zeal for the propagation of his chosen belief system, that the message was officially carried to Sri Lanka where the religion developed in a way unique to the island. By that time, Buddhism had undergone some changes. Upon the Buddha's death, the first of four Buddhist councils had been held in Rajagriha. Tradition has it that both the rules of the Order and the Buddha's sermons on matters of doctrine and ethics were recited by disciples at the first meeting.

A century later, differences within the Sangha, or Order of Monks, had become so marked that at a second council in Vaisali the Sangha is said to have split into two sections. While the points of controversy were still relatively minor, these would be followed by doctrinal differences of greater magnitude. The schism created the 'Believers in the Teaching of the Elders' or Theravadi, and the 'Members of the Great Community' or Mahasanghikas. A third great council took place in Pataliputra during the time of Ashoka, although it is doubtful that the Emperor himself actively participated in it. The differences within the Sangha resulted in the expulsion of many heretics and the establishment of the Theravada school as the orthodox. This paved the way for the later schism of Buddhism into the Little Vehicle (Hinayana) and the Greater Vehicle (Mahayana) branches. The former has been said to teach the attainment of salvation for oneself alone, the latter the attainment of Supreme Buddhahood and the salvation of all. Later, taking the example of the Emperor Ashoka, King Kanishka convened the fourth great Buddhist council in Kashmir to settle disputes of faith and practice. At this meeting the Theravada doctrines would be codified.

Under the Emperor Ashoka's patronage, Buddhist missionaries were sent to several areas of the subcontinent. The purpose of these missions was not only to preach Buddhism but to establish the Sangha locally so as to ensure the perpetuation of the religion. One such mission went forth to Sri Lanka during the reign there of Devanampiya Tissa. It was to be Ashoka's most successful campaign. The general level of acceptance was heightened by the conversion and royal encouragement of the Lankan king himself. Ruler of Anuradhapura, initially a village founded three generations after Vijaya's day, he was initiating the process of increasing the authority of his kingdom over other areas of the island. Buddhism would prove to be a powerful element of unification, culturally and politically.

The mission was considered of special importance, for it was led by Mahinda, either son or brother of the Emperor. Friendly contact between the Mauryan court and that of the recently enthroned Devanampiya Tissa had begun already before the mission, and the Emperor himself had supported the former's royal consecration. The receptiveness of the court to Buddhism strengthened the bonds of friendship between the Kingdom of Anuradhapura and the Mauryan Empire. Gifts and envoys were exchanged frequently.

13. Bodhi tree in stone relief found in Ayuddhya, Bangkok National Museum. (Luca Tettoni/Photobank)

Among the most memorable and enduring of these was a cutting—the selection having been supervised by the Emperor Ashoka himself—from the bodhi tree under which the Buddha had sat in Bodhgaya. It was planted in Anuradhapura where the ensuing tree still stands. The branch was brought to Sri Lanka in the care of Mahinda's sister, the nun Sanghamitra. Following Mahinda's initial success, Sanghamitra was called upon to establish an Order of Nuns so that King Devanampiya Tissa's sister-in-law and other women might be ordained. Other sacred relics, such as the alms bowl of the Buddha, and one of his teeth, would follow in later years. An annual festival in honour of the Tooth Relic is still held today. It is of interest to note here another ritual: the festival of the lustration of the Buddha image, mentioned in a tenth-century inscription.

The royal interest in Buddhism meant intimate links between the state and religion. Indeed, it was with the embrace of Buddhism that the great architectural history of the island began as well. The King granted a royal park as a residence for the ordained priesthood and this was the beginning of the Mahavihara, the great body of monks which would become the centre of Buddhist orthodoxy. Through the efforts of Devanampiya Tissa and his successors, Anuradhapura itself would be transformed into a sacred city, studded with palaces and mansions, monastic buildings (including a nine-storey residence for the Mahavihara), and stupas of a size never before seen.

Over the centuries following the introduction of Buddhism, religion and royal authority usually supported each other, drawing strength from their alliance despite the political instability which would haunt successive dynasties. Then, too, pressure from and clashes between South India and the Sinhalese alternated with periods of harmony and friendly exchange. The proximity of India to Sri Lanka meant that there had long been close contact between the two countries. While initial settlers had heralded from Aryan India, it was through the vicissitudes of its political and cultural intercourse with the people of the south that the Sinhalese nation grew into a distinctly separate entity, and Buddhism was a motivating factor in this quest for national identity. If Vijaya was the founding father of the Sinhalese race, and Devanampiya Tissa the king who established the Dharma on the island, it was King Dutthagamani in the second century BC, a figure fused of history and religious chronicle, who united the Sinhalese under the banner of Buddhism. Kings went on to extend their patronage to the adopted faith. Hundreds of thousands of men and women of the laity embraced it; thousands entered the Sangha.

It was the teachings of the Theravada order which Ashoka's missionaries had brought with them. Despite the intermittent influence of Mahayana Buddhism, the Theravada sect took deepest root. The Theravadans were conservative in their observance of disciplinary rules and in their approach toward the doctrines. The Dharma was memorized and relayed orally, as it had been since the days of the Buddha. But, shortly before the turn of the new millennium, in the reign of Vattagamani Abhaya, the *Tripitaka* was at last committed to script. That this milestone took place in Sri Lanka would influence international relations with key countries in the region. Generations of monks came to seek the purest available source of the sacred Canon.

Anuradhapura, despite its long pre-eminence, was not a highly centralized autocratic structure but one in which the interests and identities of outlying regions had to be tolerated. While this in some ways weakened royal authority, it also permitted life at village level to flow on undisturbed, even during times of political turmoil. This was aided in no small measure by the island's sophisticated irrigation system.

It is probable that the Indo-Aryans settlers brought with them the techniques for rice cultivation and irrigation. But it was their descendants who raised this basic knowledge to a skill of intricate magnitude. While the earliest projects were devoted mainly to conserving rather than diverting water, by the third century AD vast areas of the island were fed by lengthy canals whose sources were colossal reservoirs and dams.

The results of these splendid irrigation works were several. The taming of the land often meant agricultural surplus and a general spreading out of the population. The fact that ownership was not limited to the state meant a sustained interest in productivity by the land-owning monasteries and private individuals. As the complexity of the irrigation systems advanced over the centuries, so too did trade. Surplus grain was an important feature of indigenous trade, and seed was even used as a form of banking currency. Of greater interest to foreign traders, from ports north, east, and to the west, were the lures of pearls and gemstones, ivory and elephants. However, despite increasing revenues for the state, the island's early economy was not based on trade.

The agricultural surplus, through the evolving systems of levies on land and water, made possible enormous investments in the architecture and sculpture of the Anuradhapura Kingdom. The first concepts of Buddhist art came on the winds of the Mauryan mission. Thus, the earliest Buddhist monuments in Sri Lanka were rooted in the building traditions of the Buddha's homeland. But just as the very concepts of the religion were transformed in their embrace by the peoples of distant lands and passing time, so too did the artistic styles metamorphose into those distinctive of their adopted homelands. The essence of India remained, but the life's breath was in the rhythm of the new country.

The characteristic monument of early Buddhism was the stupa, which had its roots even in pre-Buddhist days as a burial mound. Enshrining relics of the Buddha, or other objects of veneration, they rose, solid and serene, to dominate the emerald panorama of the island. The earliest of the important stupas at Anuradhapura was built in the reign of Dutthagamani. The most colossal, larger than any in India, was built during the reign of King Mahasena in the late third century AD. Of ornamental importance were the stupas' decorative frontispieces. The technique of stone-carving, said to have arrived with Prince Mahinda, would find some of its finest expression in the creation of moonstones, large, richly decorated semicircular stone slabs.

Along with ideas, images of the Buddha had journeyed to the island from India. It was the Amaravati style of Andhra in the south-east that shaped the earliest arrivals in the first century AD.

14. Stupa, Ruwanweli Dagoba, Anuradhapura. (Luca Tettoni/Photobank)

However, the flavour of the island itself soon found expression. Many of the images later found on foreign shores to the east could be seen as being of characteristically Sri Lankan craftsmanship, of which one feature could be said to be a certain restraint. The commonest pose of the island's seated images of the Buddha is that of *virasana*, the position which the Emerald Buddha shares. Indeed, based on some of its features, the possibility of a Sri Lankan origin has been proposed for the statue. Given the island's position in the maritime trade of the region, it would not be beyond the realms of the unthinkable that a large piece of jade could have found its way to those shores, to be reverently crafted by expert Sinhalese hands.

The cultural and religious ties with India would continue to manifest themselves in the indigenous sculpture, so that aspects of Indian influence in its manifestations of various schools appeared in the sculpture of Anuradhapura for centuries. And once the art of stone-cutting had been mastered, the Sinhalese

Buddhas, like their stupas, loomed in sizes greater than those in India as colossal images were carved into rock faces.

During the early centuries of the first millennium AD, around the time when the Chronicle claims the Emerald Buddha set out on its international travels, indigenous power struggles occupied the rulers of Anuradhapura. But in the late fifth and early sixth centuries, during the reign of Moggallana, South Indian mercenaries were brought in to assist in a succession dispute. This

15. Giant Buddhas, Gal Vihara, Polonnaruva. (Luca Tettoni/Photobank)

16. Vatadage, moonstone in the foreground, Polonnaruva. (Luca Tettoni/Photobank)

17. Probably King Parakramabahu I, Polonnaruva. (Bangkok National Museum)

opened the door to Tamil influence in the Sinhalese court and South India would be reluctant to close it again. There followed centuries of political tightrope walking as Sri Lanka sought to avoid the pitfalls of South Indian hegemony.

The rise of Hindu kingdoms in these centuries—the Pandyas, Pallavas, and Cholas—put religion squarely into the power picture. By this time in India, and especially in the south, adherents of Buddhism had declined drastically in number, causing antagonism of a religious nature to join that of ethnic.

But not all invasions were from India into Sri Lanka. At least twice Sinhalese armies invaded South Indian kingdoms, seeking to support rival Indian rulers in a bid to keep the most powerful away from the island. Ultimately, the South Indian forces proved too strong, and in the early eleventh century the Cholas, during the reign of Rajaraja the Great, were the ones to bring the Sinhalese under their rule. For security reasons, the capital would be shifted from ancient Anuradhapura to Polonnaruva.

Not until a century later would King Vijayabahu I be able to take advantage of the Cholas' difficulties in South India to shake off the yoke of foreign domination. He and King Parakramabahu I of the mid-twelfth century were instrumental in bringing about the recovery and revitalization of Buddhism on the island. The unification of the Sangha, and the architectural activity at Polonnaruva effected by the latter monarch would be a fitting victory to the efforts begun by King Vijayabahu. It is this king of the eleventh century who was a contemporary of Pagan's King Anawrahta (Anuruddha of the Chronicle) and thus, despite a time warp of half a millennium, we once more rejoin the pages of history and those of the Chronicle of the Emerald Buddha.

4
Burma

Being a continuation of the Chronicle, followed by a brief look at the land known as Burma, taking note of a few historical and religious connections between Sri Lanka and Burma.

THE Chronicle continues: informed of their intent, the Lankan king welcomed the visitors with feasting and gifts. Silakhandha soon made his way to the island's patriarch, explaining his doubts that the Indian texts were suitable for conducting ordination ceremonies. Thus, not only were copies of the Lankan texts sought, but the Pagan priests requested permission to take orders again. Much rejoicing by the inhabitants accompanied this event. Some time thereafter, the transcribing of the *Tripitaka* and Commentaries was accomplished as well. The Siamese version of the Chronicle states that the priests and the Scriptures written by the Lankans were placed in one of the boats bound for Pagan while those written by the people of Pagan, along with the Emerald Buddha, went into the other. But the winds of fate were blowing, and the vessel carrying the statue was to take a different direction from that of human desires.

Before following the Emerald Buddha to yet another distant land, let us pause to shine a faint light across the backdrop of the historical stage. Recorded history pays homage to King Anawrahta as the legendary and historical ruler who united peoples and places under the banner of a Theravada Buddhist Burma. The fact that this monarch ruled some six hundred years after the date attributed to him in the Chronicle may, in the spirit of this voyage, be laid aside. Despite the tradition that Buddhagosa brought Pali commentaries and canonical literature to Thaton in AD 403 from Sri Lanka, the earliest recorded religious contacts between Burma and Sri Lanka date to the eleventh century. Indeed, the priest known in the Chronicle as

Silakhandha may well be the Mon monk Shin Arahan who, tradition says, worked under King Anawrahta to convert the Burmese to Theravada Buddhism.

It was the Chinese who first mention Burma, in a 128 BC reference to the old overland route linking China and the West. But until the eleventh century, documented records as to the country's history are few. The earliest of Burma's people remain nameless. Then, with roots reaching deeper than the written word, came the principal populations. The Pyu would practise Hinduism and both types of Buddhism. Their great capital at Srikshetra left behind the earliest known sculptures, displaying connections with both the Pallava and Gupta styles of India. The Pyu Kingdom disappeared in the ninth century. Enduring longer before their assimilation, the Mon of Burma were close relatives of the Eastern Mon, whose political and cultural centre, Dvaravati, flourished from the sixth century on. The Western Mon had their capital at Thaton, a thriving centre of Theravada Buddhism. Next to arrive was the main body of the Burmese people, coming down from the Chinese–Tibetan border in the ninth century. By the time of King Anawrahta, the Burmese were already in the process of being converted to Buddhism although the worship of *nat*, a collection of nature spirits, ghosts, and deities, would continue to play an intrinsic role in both the religion and art of the country to the present day. The Shans would come after the Burmese, acquiring principalities in Upper Burma.

Just how Buddhism originally arrived in Burma remains unclear. Whether it had travelled the overland route with Indian merchants, settlers, or missionaries, or come by sea, the Buddhism at the time of King Anawrahta was of both the Mahayana and Theravada varieties. Burmese Buddhist legends have their own versions of religious milestones. One of the earliest concerns two Indian brothers to whom the Buddha gave eight hairs from his head. The brothers are said to have brought them to Suvarnabhumi, in this case thought to be Lower Burma, the Golden Land referred to in the *Jataka* tales. A Mon chronicle relates how two of Emperor Ashoka's missionaries, Sona and Uttara, dispatched by the Third Buddhist Council, arrived in Suvarnabhumi to revive Buddhism. Whatever the contacts between India and Burma, their subsequent influence on Burmese architecture, sculpture, and philosophy of life was strongest in its Buddhist manifestations.

By the time of King Anawrahta, however, Burma found itself looking toward Sri Lanka for support in religious matters. Hinduism had triumphed in India's once great Buddhist centres. It is said that the King, informed by the monk Shin Arahan of the availability of the *Tripitaka* in Thaton, invaded that city because its ruler would not willingly part with the copies. The result for the Pagan king was possession of both the *Tripitaka* and Lower Burma. A Burmese chronicle further relates how King Anawrahta subsequently sent four warriors to Sri Lanka to bring back the *Tripitaka* for comparison by Shin Arahan with the scriptures taken from Thaton.

Relations between Burma and Sri Lanka were on a friendly footing during King Anawrahta's reign; commerce and politics as well as religion were factors in this. In the eleventh century, King Vijayabahu of Sri Lanka had risen against the Chola occupation. But unlike previous political struggles, wherein the island's warring rulers might have turned to South India for aid, in this case the Cholas were too dominant on land and sea for any of the other South Indian kingdoms to offer any real assistance. Burma, however, was a growing power in the region, its dominions stretching to the north and southwards. Anawrahta's conquest of Mon Thaton was a major milestone for Burmese civilization. Sri Lankan Buddhist monks who had fled to Burma during the Chola occupation may have sent glowing reports on the power of the Burmese king, motivating Vijayabahu to seek support from his royal peer.

Whether or not the Burmese actually did send military aid along with economic assistance to help in the successful expulsion of the Cholas, Anawrahta had incentive to ally himself with the island king. The former's push for control of the Kra Isthmus, which through the Chola invasion had weakened the authority of the Srivijaya Empire in the Malay Peninsula, meant that any

1. A view of Wat Phra Keo in the compound of the Grand Palace. (Luca Tettoni/Photobank)

2. Interior of the Chapel Royal. (Luca Tettoni/Photobank)

3. Inner view of Sanchi North Gate. (Bangkok National Museum)

5. Buddha in Cave No. 1, Ajanta. (Luca Tettoni/Photobank)

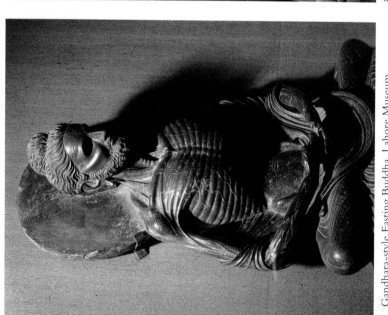

4. Gandhara-style Fasting Buddha, Lahore Museum. (Luca Tettoni/Photobank)

6. Remains of the Asokarama Monastery, with moonstone in the foreground, Anuradhapura. (Luca Tettoni/Photobank)

7. Procession of the Tooth Relic, Kandy. (Bangkok National Museum)

8. Degaldoruva fresco, Kandy. (Luca Tettoni/Photobank)

9. A shrine on the platform of the Shwe Dagon Pagoda, Rangoon. (Kelly, 1933)

10. 'Four-faced towers, Bayon, Angkor Thom. (Luca Tettoni/Photobank)

11. Khmer King Jayavarman VII, found at Phimai, Bangkok National Museum. (Luca Tettoni/Photobank)

12. General view of Angkor Wat. (Luca Tettoni/Photobank)

13. Detail of a frieze, Bayon, Angkor Thom. (Luca Tettoni/Photobank)

14. Lanna-style Buddha, Bangkok National Museum. (Luca Tettoni/Photobank)

15. Aerial view of Sukothai. (Luca Tettoni/Photobank)

16. Ayuddhya, Dutch oil-painting, c.1650. (Rijksmuseum, Amsterdam)

17. Wat Phra Keo, Chiang Rai. (Luca Tettoni/ Photobank)

18. Ayuddhya-style Buddha, Bangkok National Museum. (Luca Tettoni/ Photobank)

19. Wat Phra Keo Dontao, Lampang. (Luca Tettoni/Photobank)

20. Ho Phra Kheo, rebuilt in the nineteenth century, Vientiane. (Steve Van Beek)

21. Monks in Wat Mai, Luang Prabang. (Steve Van Beek)

22. Royal barge on the Chao Phya River. (Luca Tettoni/Photobank)

23. Mural, Wat Phra Keo. (Luca Tettoni/Photobank)

24. Thai lacquered panel. (Luca Tettoni/Photobank)

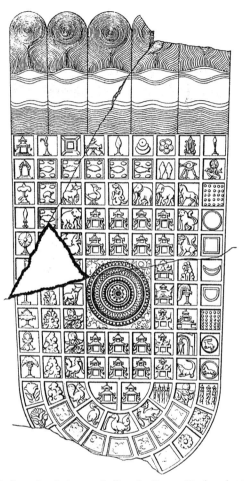

18. Buddha's footprint, Loka-nanda Pagoda, Pagan. (Archaeological Department of Burma)

lessening of Chola power would strengthen his own hold on the area. Then, too, was the fact that unlike the Cholas, the Sinhalese and the Burmese shared a common faith.

Aside from some unsettling interruptions of a political nature, such as a battle arising thirty years later over King Alaungsithu's

19. Kubyaukkyi murals, Pagan. (Luca Tettoni/Photobank)

worries about the rapidly expanding Kingdom of Cambodia and his fears of ties between that neighbour and Sri Lanka's King Parakramabahu I, close religious contacts were maintained between the two countries for centuries. Its Order re-established, Sri Lanka was becoming the major source for religious revival; Burmese monks looked to the fraternities of the Sinhalese Order wherein Theravada Buddhism was believed to be the closest to its original form. The Tooth Relic enshrined in Pagan's Shwe Dagon, the Chinese-inspired pagoda begun by Anawrahta in 1059, is a duplicate of that in Sri Lanka, perhaps a gift of thanks from the island's ruler to Anawrahta.

It is thus possible, through the tangle of facts, tradition, suppositions, and supernatural feats, to view this segment of the Chronicle of the Emerald Buddha as illustrative of the countries' friendly ties during the days of King Anawrahta and his—in the Chronicle—unnamed contemporary Lankan sovereign.

5
Cambodia

Being a continuation of the Chronicle, as well as an observation of the roots of the Khmer Empire of Cambodia and some curiosities of Angkor.

THE Chronicle continues: King Anuruddha, who had supernaturally flown back to Pagan, grew concerned about the length of time it was taking for the boat with the Emerald Buddha and the transcription of the *Tripitaka* to reach his kingdom. Informed that the statue had arrived in Indapatha Nagara (Angkor), the king flew to that town. There he pretended to be but a messenger of King Anuruddha, telling a priest in a monastery to inform his king that he had been commanded to ask for the return of the boat and scriptures. The King of Indapatha Nagara, pleased at the unexpected treasures, refused.

Piously holding back his murderous wrath to avoid sinning, King Anuruddha chose to induce the King of Indapatha Nagara and his men through a display of a brush with death. He rode through the air, leaving the mark of a wooden sword upon the throats of the royal entourage. At the same time, he threatened to cut off their heads should the *Tripitaka* not be returned by the following day. Convinced of the potential reality of this, the King hastily sent for the captured boat. Yet, King Anuruddha took only the vessel and the *Tripitaka*, leaving behind the Emerald Buddha. The Pagan king is credited with the spread of Buddhism through the correctness and purity of the *Tripitaka* and Commentaries.

The sovereign of Indapatha Nagara and his people showed great reverence to the Emerald Buddha, and while the image resided there Buddhism flourished in the town. However, during the reign of Senaraja there occurred an incident which was to have grave consequences: the young son of an officer had a spider which ate the pet fly of the King's own child. Whether out of paternal rage, or fearing a power struggle for his heir in the distant future, the King ordered the spider's owner drowned. Thereupon a *naga*, or multi-headed serpent, residing in the great lake caused a hurricane and flood, and the King and almost all inhabitants of the town perished. A concerned priest took the Emerald Buddha to the north.

It is only fitting that the Chronicle of the Emerald Buddha carries the image to Cambodia. The journey would have been first by sea, and then probably overland across the Isthmus of Kra to the east coast of southern Thailand. From there a ship would have crossed the Gulf of Siam, to the land of the Khmers. Theirs was an empire whose builders had visions of greatness, whose political and cultural prestige transcended man-made boundaries. Born of two kingdoms deeply rooted in the spiritual traditions of India, the Khmer Empire would rise to become, for some centuries, the most powerful state in Indo-China.

The beginnings of Cambodia are found earliest in the Kingdom of Funan, whose territories initially occupied the lower valley of the Mekhong River. A third-century Chinese account of a local legend traces its origins to the marriage of a Brahman named Kaudinya and Soma, a native princess who was half human and half *naga*. While this version may be based on a similar royally inspired legend of the South Indian Pallavas, its existence here suggests that already from early on relations between Cambodia, India, and China were an important element in Khmer culture. Two practices of Funan, in particular, would be adopted by the Khmer kings of Cambodia: the cults of the *naga* and the sacred mountain. The symbol of the nine-headed serpent would become dominant in Khmer iconography. The concept of the sacred mountain could well have had roots reaching deeper even than ancient India, to the ziggurats of Ur. Funan adopted the cult of temple-mountains, as would the Sailendras in Java; the Khmers, too, continued the practice.

Funan's geographical location made it ideal as a trade centre. Ships carrying goods between southern China, the east coast of India, and the world beyond sailed into its ports. The influence

20. Port of Chantaboun. (Mouhot, 1864)

of India would be the strongest. Contact with Indians from Conjevaram in the south, Tamralipti at the mouth of the Ganges, and Broach and Cragnore on the west coast, amongst others, contributed to the gradual spreading of Indian culture to the east. It must initially have been through trade and concomitant trading communities, not formal colonization, that the Indian influence began to spread, already in the years before the opening of the Christian era. Hinduism, no proselytizing religion, could have been absorbed through communities of Hindus settled in the region.

That these communities were permitted to reside points to the probability that Indian culture, religion, and social structure were accepted by the highest echelons of native society. It is likely that local rulers, seeking to increase their prestige, called in Brahmans as advisers; Funan was organized along the lines of an Indian kingdom. Thus, while trade may have instigated contacts, the lasting cultural elements were transmitted through the courts. At the same time, and equally important, inhabitants of these lands travelled westward to India, returning with Indian traditions.

Local kings followed Kaudinya. During the reign of Fan Shihman in the third century, the Funan frontiers moved further east, southwards, and to the west. Although the rulers of Funan practised Saiva Hinduism, with its pre-Buddhist memories of phallic worship, both Hinduism and Buddhism were allowed to flourish in the kingdom. Most of the Indian cults of the time are said to have existed in Funan. The kings of Funan also drew from successful technical concepts in Indian city planning and waterworks. A system of canals, upon which boats could sail through moated cities, allowed for irrigation and flood control of the Mekhong River.

By the middle of the fourth century, an Indian was once again on the throne. Brahmans and other kings with Indian names followed. And little more than a century later, Kaudinya Jayavarman cultivated good relations with China, sending Funanese monks to China to translate Buddhist texts from Sanskrit into Chinese.

21. A lingam at Preah Khan. (MacDonald, 1958)

Rudravarman, a devotee of Vishnu, was the last great king of Funan before its conquest by the neighbouring Indianized Kingdom of Chenla, a former vassal kingdom of Funan. While Funan stretched over today's southern Cambodia and Cochin China, Chenla lay to the north, occupying what is now northern

Cambodia and southern Laos. Chenla, too, was born of myth, the union of a hermit, Kambu Svayambhuva, with the celestial nymph Mera, given in marriage by the god Shiva. The houses of Funan and Chenla are said to have been joined later in a matrimonial alliance. Although Funan would continue to emit life signs into the seventh century, by 627, in the reign of Isanavarman I, it was incorporated with Chenla. Under this same king, power was extended towards the area that would later become the centre of the Angkor monarchy as well as to the eastern borders of the Mon Kingdom of Dvaravati.

With the annexation came the greatest phase in the foundation of classic Khmer art. The Buddha images of Funan recall the styles of Amaravati, Sarnath, and Ajanta. But while India continued to provide the blueprints, the evolving styles were not mere copies of Indian works but the results of fusion with existing, masterly indigenous artistic traditions. With the annexation, too, Funan sculptors were able to make use of the stones of Chenla, and sculpture was the major art form during the Funan Chenla era.

Different forms of Hinduism were royally endorsed, and kings derived royal authority from a deity; sculptures of the Hindu gods and goddesses were many. By the time of Bhavavarman II in the mid-seventh century, however, Mahayana Buddhism was spreading into the kingdom. Similar to their royal peers in northwest India, as claim to their sovereignty, local kings made use of the images of Bodhisattvas.

Just as within fabrics the warp and weft from time to time display snags in the weaving, although the overall pattern succeeds and the finished work is a masterpiece, so too did the Kingdom of Chenla fray, leaving other threads to pull the design into place. In the late eighth century, the Sailendra Dynasty of Java, following the Kingdom of Srivijaya in Sumatra, closed the curtain on Chenla. Earlier rulers of Java had been worshippers of Shiva and Vishnu, and Hinayana Buddhism had flourished for some time, but the Sailendras of Java, creators of the great monument of Borobudur, were followers of the Mahayana and Tantric Vajrayana forms of Buddhism.

22. Borobudur. (Embassy of Indonesia, Bangkok)

Because of the increase in trading vessels moving between China and India, the Kingdom of Srivijaya had prospered. This, in turn, had caused the Javanese kings to cast an eager eye to the north-west, and the Sailendra's connection with Cambodia grew with the increase in the dynasty's power which was felt in Malaya and Tonkin as well. Perhaps harbouring descendants of those who had fled to Java during the fusing of the two pre-Angkor kingdoms, the Sailendra Dynasty claimed kinship with the ruling house of Funan. Around 790, a young man who had lived a good part of his life at the Sailendra court returned to the land of his birthright to liberate it from Javanese suzerainty. Despite desiring this element of separation, Jayavarman II was not adverse to emulating the Sailendras' splendid trappings of dynastic power in his establishment of the Angkor Kingdom.

In the year 802, he called upon a Brahman to preside over a ceremony of consecration. A Shiva lingam, symbolizing his royal omnipotence, was set upon a hill, Phnom Kulen, it in turn symbolizing Mount Meru, the mountain dwelling-place of the gods. The ceremony marked the birth of a new era, the beginning of the Khmer Dynasty. Jayavarman moved the site of his capital city several times. The first large building projects began during his reign, with artists imported from as far away as Java and Champa, today's Vietnam.

If the seeds of Khmer art were planted during Jayavarman's time, it was under Indravarman I that they burst into full bloom. A scholar as well as a ruler, this king of the late ninth century brought peace to the Khmer Kingdom of Cambodia and laid the foundations of Angkor, undertaking irrigation schemes which increased rice harvest and thus the urban populations. The initial concept of irrigation, probably imported from India and made use of by the citizens of Funan, was expanded upon. The link of heavenly water and prosperity meant that Angkor was regarded as divine. Royal support and proper ritual, Brahman and later Buddhist as well, were integral aspects. Thus, religion and royalty went hand in hand, as attested to by the dedicatory statues, temples, and palaces which were associated with the cult of king-worship. To begin with, hilltop sites were chosen for religious edifices. Later, the buildings themselves featured towers marking them as temple-mountains. Indravarman's own was the Bakong, inspired by Borobudur.

During the centuries and succession of Khmer kings which followed, emphasis was placed on the continuing architectural and economically based splendours of Angkor. Yasovarman I opened up new lands for cultivation through new chanelling of waters. His own temple was the Bakheng, atop Phnom Bakheng, and the mystique of Borobudur was also mirrored in its conception. During Rajendravarman's reign, several Mahayana Buddhist establishments were set up at Angkor. The concept of his temple-mountain becoming a king's funerary shrine probably became established at this time. A vibrant work of this period was the temple of Banteay Srei, built during the reign of Jayavarman V and founded by the King's adviser, again a Brahman.

Shortly after the dawn of the eleventh century, following an internal Angkorian power play over the formerly independent Dvaravati centre of Lopburi, subsequently a province of Angkor, a new ruler came to the throne. This was Suryavarman I, contemporary of Anawrahta, creator of the palace and grand plaza at Angkor. With this king, too, came the golden age of Angkor. Although in the Chronicle of the Emerald Buddha no name is given to the ruler of Indapatha Nagara at the time of the image's arrival there, Suryavarman I, and the two sons who succeeded him, Kings Udayadityavarman II and Harshavarman III, were all contemporaries of Anawrahta. Any one of them could have been reigning at the time the Emerald Buddha arrived from its long journey across sea and land. In which royally appointed chamber would the image have been ensconced: the Ta Keo, the first temple-mountain made entirely of sandstone, the imperial palace of the Phimeanakas, or the Baphuon? The legendary arrival of the Emerald Buddha in Angkor assuredly heralded an era of good fortune.

Indeed, for the Khmers the eleventh century was a time of political and cultural flourishing. The effects were felt in varying degrees by the people of the cities, towns, and countryside as

23. Facade of the Temple of Angkor Wat. (Mouhot, 1864)

well, for the Khmers controlled their empire extensively, even outside the nucleus of the kingdom. Governors and princes looked after the provinces, such as Phimai, Sawankhalok, Ratburi, names which today are to be found within the confines of modern Thailand. Although himself a Hindu, Suryavarman I tolerated and even assisted the spread of Mahayana Buddhism. Grandiose monuments, each greater than those before, required vast funds and manpower; both were available to the rulers of Angkor. The architects and sculptors of Suryavarman II, who reigned between 1113 and 1150, were undoubtedly aware of the drama of design. Angkor Wat can be said to mark the pinnacle of Khmer art, and within the structure can be seen the evolution of its spirit: the brilliant fusion of Hindu and Buddhist concepts and king-worship, a breathtaking mixture of the simple and sophisticated.

Yet, the spotlight of glory was fading for the Khmers, and that player which had featured so often in the royal dramas of the ancient Greeks would have its turn upon the Khmer stage. The Angkor kings, intent upon demonstrating divine royalty and a collection of architectural splendours, allowed hubris into the cast. Neglecting to nourish the very machinery which kept them in place, successive kings lost sight of priorities. Irrigation canals were left to silt up while subjects raised the monuments to their god-kings. Although Suryavarman II did expand the borders of the empire to its most wide-reaching ever, his mind could not have been on the more delicate aspects of political relations. In 1145, annoyed at Champa's refusal to assist in his efforts to control Annam, he simply deposed the king. The Chams bided their time, and upon Suryavarman II's death during another attempt to take over Annam, Angkor experienced its first invasion. Sailing up the Mekhong into Angkor, the Chams burned the wooden city to the ground in 1177, carrying away with them treasures which were centuries old. Did the invaders demand the Emerald Buddha? Assuming the image was there, it must have been well hidden, and one can imagine the suspense as the looters swept through the smoke-filled city, searching for prizes

24. Carvings of arms, utensils, and ornaments at Angkor Wat. (Mouhot, 1864)

to take home with them. But, for the Emerald Buddha, Champa was not to be.

As if to prove that Angkor would not be vanquished, four years later Jayavarman VII took to creating new cities and temples, and extending the boundaries of the empire to the northwest, north, and south. Driving out the Chams, he proceeded to make Champa yet another province of the Khmers. Jayavarman's building projects outweighed any of his predecessors, in terms of quantity if not stylistic quality. His reign marked a religious milestone. Due perhaps to a fear that the Hindu pantheon had abandoned Angkor, Buddhism, which had been gaining ground amongst the common people, became the official religion of the king himself although Brahmanism retained its hold on the royal court.

The thirteenth century began with the construction of Angkor Thom, the city built over the site of Udayadityavarman II's Baphuon. Jayavarman VII's temple-mountain, the Bayon, rose out of the centre of Angkor Thom. Its inspiration was Buddhist in nature; the worship of the Bodhisattva featured in its adornments. There, too, the old iconograph of the Khmers, the cosmic *naga*, was combined in sandstone with the *naga* who in Buddhist tales sheltered the Buddha from the stormy waters of Mara, or Death, which sought to impede the Buddha from attaining Enlightenment.

Although several kings followed Jayavarman VII, the splendour of Angkor and its socio-economic power were on the wane. The borders shrank with the coming of the Siamese. At the same time, these invaders strengthened the spread of Theravada Buddhism. For despite the royal favour of Mahayana Buddhism, Theravadism had been gaining acceptance. Political contacts between Parakramabahu I's twelfth-century Sri Lanka and Cambodia had probably made many aware of the religious activities and importance of the island. Indeed, it is possible that a son of Jayavarman VII may have gone on a journey there with a group of Burmese monks, thereafter spending his life promoting Theravada Buddhism as a monk in Pagan.

As mentioned in Chapter 3, Burma had developed strong

25. A *naga* head at Preah Pulilay. (MacDonald, 1958)

religious links with Sri Lanka from the eleventh century onwards. At the same time, Theravadism had long been important in Dvaravati, and much of today's Thailand shared the Theravada belief. When Ramkhamhaeng of Sukothai expanded his dominions by shrinking those of the Khmer Empire in the thirteenth century, Theravada Buddhism was a natural product. The seeds of the faith would have fallen on fertile ground, for the simplicity of Theravada Buddhism must have appealed to a people who for generations had lived in the shadow of the temple-mountains.

Towards the middle of the fourteenth century, Jayavarman Pararmesvara actively encouraged the spread of Theravadism in the Angkor Kingdom. It was perhaps by his command that Sanskrit now gave way to Pali as the official language. Through the assistance of his daughter, Theravada Buddhism was passed on to Laos.

In 1431, the death knoll of Angkor sounded. Tiring of decades of skirmishes between the Khmers and themselves, Siamese forces under King Boromaraja II of Ayuddhya took Angkor, destroying the reservoirs and irrigation canals that had made life in Angkor so prosperous. And although the Khmer court continued in Phnom Penh, the glories of the Khmer civilization would not be repeated.

Whether or not the ruler in the Chronicle of the Emerald Buddha who commanded the cruel punishment in the incident of the pet fly and spider, causing the subsequent destruction of the town by the *naga*, was actually the Khmer King Dharmasoka whose death during the siege of Angkor in 1431 was followed by the defection of two leading monks and mandarins to the Siamese, thereby aiding the collapse of the city, the past and the Chronicle overlap here: Angkor was abandoned. The Emerald Buddha was gone.

6
Conundrum

Being a continuation of the Chronicle as well as a glimpse of the conundrum thereof.

THE Chronicle continues, with increasing contradictions and confusion of time and place: a priest spirited the image away from Angkor to a northern village. Thereafter, a king who was reigning over Ayuddhya took possession of the Emerald Buddha as well as many of its guardians. The inhabitants of his land were followers of the Three Gems, and ready to honour the image. Likewise, people came from many directions to make offerings and, for a time, the Emerald Buddha stayed in Ayuddhya. A king of Kamphaeng Phet [Kampeng Bheja] then carried the image with him to his hometown where the citizens, like those in Ayuddhya, were of the faith. Again, the Emerald Buddha stayed amongst the believers except—says the Siamese version of the Chronicle—for a period of one year and nine months when a prince of Kamphaeng Phet requested of his father that he be allowed to worship and care for the image in Lopburi.

Then another prince (said in the same version to have taken the Phra Sihing, another revered and also much coveted Sinhalese image of the Buddha to Chiang Rai) now took the Emerald Buddha to that town. A prince of Chiang Mai, Keu Na, then asked that the image be brought to him in his city, where a pavilion had been built for it in his palace. The Emerald Buddha remained there until 1506 [*sic*], revered by the citizens and honoured in annual festivals, and the religion of the Buddha flourished in Chiang Mai. The Chiang Mai version of the Chronicle explains that it was during the reign of Sen Muang Ma in the years 1388–1411 that the image was hidden behind stucco in a Chiang Rai temple. It is said that no one knew what it was, but that danger came to those who touched it.

Another version, however, that of Phra Dhatu Chom T'ong, claims it was the priest who took it from Angkor to Chiang Rai. There, without saying a word to anyone, he constructed a monastery near the Me Kok River. The Emerald Buddha was placed inside and, to safeguard it, covered with a coat of stucco. When Muna Gama was chief of Chiang Rai, he ordered the monastery restored, and it was then that the stucco was broken and the statue revealed.

Historians credit this event as having occurred in 1434, during the reign of Sam Fang Kaen. If one follows the

26. A high priest of Bangkok. (Bock, 1884)

Phra Dhatu Chong T'ong version of the Chronicle, it would not be impossible that the statue, rescued from Angkor during its destruction in 1432, had spent the interim in Chiang Rai. But why would a new monastery need to be restored after only two years? Perhaps the threads of the tapestry unravel in different directions: that the destruction of Angkor referred to in the Chronicle was not that by Siamese hands in 1431, but the earlier siege of the Chams, in 1177. If this were the case, then the travels of the Emerald Buddha from Angkor to Ayuddhya, Kamphaeng Phet, and Chiang Rai would be plausible.

However, King Ku Na of Lan Na, referred to above as Keu Na, reigned only between the years 1355 and 1385. Then again, the King Ditta whom the Ruang Pongsawadan Yonok version of the Chronicle speculates was the Ayuddhya ruler who first lay claim to the Emerald Buddha upon its departure from Angkor, was said to have reigned even before his namesake Adittaraja of the eleventh century. This would mean that the image would have left Angkor sometime between the founding of Angkor by Jayavarman II in the early ninth century and before Adittaraja's reign. However, according to other events in the Chronicle, the Emerald Buddha was still in Sri Lanka at that time!

Clearly, the threads of history and legend are now tightly tangled. What emerges, however, is a tapestry depicting the closely linked relations between the lands and rulers of the time. Religion, commerce, culture, and politics bound the lives and fates of the people of the South-East Asian region. Ayuddhya, Kamphaeng Phet, Chiang Rai, Lampang, Chiang Mai, and Luang Prabang—all are sites mentioned in the quest for possession of the Emerald Buddha. The Chronicle here contains various versions and many contradictions of time, space, and characters. So, for now leaving the Chronicle of the Emerald Buddha at the close of the Second Epoch, the voyage herewith sets forth through clearer waters.

7
Thailand

Being an overview of some of the peoples and places that contributed to the character and composition of the kingdoms to be known in later years as Thai, as well as spotlighting a few kings and their realms at Lan Na, Sukothai, and Ayuddhya.

THE earliest of those dwellers of the valleys, lowlands, and hills of what would become modern Thailand were, like their habitats, of varied origin. Prehistoric residents of the South-East Asian region were initially hunters, developing only gradually into agriculturalists and navigators. They shared the common heritage of mankind for most of that time, but by the last centuries of the first millennium BC it is probable that their languages and cultures became differentiated. Lying beneath the dissimilitudes, however, were the dreams and memories of common traditions.

Those to whom names may first be applied are the Mon of the Menam valley and Lower Burma and the Khmer, originating in the Mekhong delta. Both would leave their mark upon the fertile fields of Thailand. Citizens of the mysterious Mon culture of Dvaravati probably lived in small city-states during the first millennium AD. Taking advantage of the trade routes crossing the Three Pagodas Pass (between the Gulf of Martaban and the Central Plain) in the sixth century, outposts of this culture eventually extended westward, eastward, and northward toward Chiang Mai and Laos. At this time, links with India were probably still vitalized through pilgrimages and the import of contemporary religious works of art. Amaravati, Gupta, and Sinhalese influences can be seen on the indigenous products of Dvaravati.

This civilization, which existed until the eleventh century, was instrumental, too, in increasing communications between regions of Thailand, such as the upper Mekhong area and the northernmost shores of the Gulf of Siam. These contacts were mainly Theravada Buddhist in nature. Connections between Lopburi, a major centre of Dvaravati civilization, and Haripunjaya, said to have been founded at Lamphun by Buddhist monks from Lopburi, remained vital for centuries. Later a reluctant province of Angkor, Lopburi would show itself a model of resistance to Khmer hegemony, strengthened by its Mon roots, and Theravada Buddhist and Tai characteristics.

For another, splintered group, thought to be immigrants from either China, the area around Dien Bien Phu in today's Vietnam, or even early inhabitants of Thailand responsible for the third- and second-century BC bronze culture of Ban Chiang in North-east Thailand, had been filtering down into what is now northern Vietnam, Laos, Burma, Thailand, and even north-eastern India. These Tai settlers organized themselves into colonizing communities which functioned as both socio-economic and political units. Survival in times of Chinese imperialism, and harmonious living with other ethnic groups meant acquiring enduring skills of politics and diplomacy.

With the advent of Angkor, the ethnic composition of the region changed. By then, the religion of the people in the area was a mixture of nature worship, Brahman rituals, Saivism, Mahayana Buddhism, and increasingly, Theravada Buddhism. At the end of the tenth century, trade between Haripunjaya and Thaton brought the Tai into the religious mainstream of Burma and Sri Lanka. While Dvaravati had been weakening from the ninth century onward, the Mon states of Burma were on the rise, especially Thaton. With the advent of Anawrahta as conquering sovereign around 1044 AD, Thaton fell under the spell of Pagan. The Burmese capital continued as a major gateway for Sinhalese ideas and religious revival. The culturally rich, Indianized civilizations of Angkor and Pagan touched the lives of even the country folk living on the further reaches of the realms.

Another power in the region for some centuries, with roots in the seventh, was Nan-chao, a Mahayana Buddhist and semi-military state located to the north in today's Yunnan province

27. Lopburi-style Buddha, Bangkok National Museum. (Luca Tettoni/Photobank)

28. Haripunjaya-style Buddha, Lamphun National Museum. (Luca Tettoni/Photobank)

of China. Nan-chao's centralized administrative control had economic and cultural effects on the uplands of South-East Asia. Three centuries later, Nan-chao and the northern Tai were linked in a quest for stability in the region, the former seeking to shield itself from Pagan might, while the latter, who had been settling in the provinces of the two great empires and upper Laos, was now interested in the open country of the lowlands.

Although Nan-chao would fade and finally crumble under the hooves of the invading armies of Kublai Khan in 1253, whose rise to power would have political repercussions on countries in the South-East Asian arena, some of the Tai groups were gaining strength. This was in no small part due to the collapse of, first, Angkor and then Pagan. Pagan hegemony was diluted by Shan moves for control in the north and Mon rebellions along the southern coasts, as well as escalating conflicts between Pagan and the Mongols.

The disintegration of the two empires resulted in the formation of smaller states and a concomitant diffusion of power. This meant that the new political entities were free to express themselves in styles best suited to their temperament and belief systems. For those in the heartland of these changes, this era would have been an age of self-discovery, of dynamic transformation before the framework of new, Tai traditions was cemented. It was at this time, too, that the Theravada Buddhism of Sri Lanka was embraced with fervour on a scale wider than ever before, with monasteries springing up throughout the region. Nakhon Si Thammarat in the south, for centuries a hotbed of power struggles among the peoples from South India, Angkor, and those of the Malay Archipelago for control of maritime trade, became a centre for Theravada Buddhism. Monks from Nakhon Si Thammarat spread the word to the far corners of the new order.

Two of those cultural centres which would influence the region were the Kingdoms of Lan Na, with its capital at Chiang Mai, and Sukothai in the north central plains. Mengrai, the man who would be king of Lan Na, hailed from the area around the future Chiang Saen. Secure in his royal background, he was quick to impose his power upon the now warring Tai factions of the surrounding regions. In 1262, while—according to the Chronicle—the Emerald Buddha resided in Angkor, Mengrai founded Chiang Rai, making it his capital. From this power base, he implemented his plan for possession of the north. Despite the probably diverse ethnic population of the region, or perhaps because of it, Mengrai was respectful of the different cultures. And although he, like many of the ruling élite of the river valley who had been influenced by centuries of Angkorian Khmer rule, was of Tai heritage, the diffuse communities of his realm would eventually blend, taking on their own identity.

King Mengrai's campaigns brought him not only territory but alliances as well. The most noteworthy were the generally harmonious relationships with his royal neighbours, Wgam Muang of Phayao and Ramkhamhaeng of Sukothai. Although

Mengrai's acquisition of the old Dvaravati centre of Haripunjaya had been one of intrigue and force, his subsequent forays were of a gentler nature, seeming to lay greater value on alliances, diplomacy, law, and the transference of craftsmanship. After forging alliances with Pegu and the Shans of Pagan, and building cities and Buddhist institutions within his northern domains, Mengrai turned his attention to the Mongols. For several years, until 1312, the Mongol armies and he engaged in struggles for supremacy in the north-east.

During this time, in 1296, the construction of the new Lan Na capital at Chiang Mai was begun. Over the centuries, this moated city would change hands several times, juggled by both Ayuddhya and Pagan as they sought to take control of the north. Despite the political vicissitudes, the sculptors of Lan Na laboured on, fashioning works in a variety of materials including bronze, wood, some stone, and semiprecious stones. Speculation as to the origin of the Emerald Buddha includes, amongst the other sites, Lan Na. Given its discovery in Chiang Rai, it is certainly convenient to attribute its creation to a divinely inspired sculptor of the region. The art of Lan Na, early on reflecting certain influences from the Pala school of that dynasty in north-eastern India and later influences from Sukothai, nevertheless celebrates a spirit that must be uniquely original.

In the latter half of the fourteenth century, during the reign of Ku Na, a well-educated king, the Sukothai monk Sumana was invited to establish a Sinhalese Order. This event would have significant effects upon the religious and cultural development of the north. Art, too, was effected. For just as Theravada Buddhism sought to express the Truth in the strict formulae of the ancient canons, so the art of Theravada Buddhism strove, through careful adherence to prescribed rules of what the image should look like, to achieve a true likeness of the Buddha.

To the south of Lan Na, another kingdom had been gaining strength and territory. An outpost established during the time of Jayavarman VII, when Angkor Thom knew glory and the Bayon breathed divinity, Sukothai was Khmer in its conception and layout. Nevertheless, toward the middle of the thirteenth

29. Wat Mahathat, Sukothai. (Luca Tettoni/Photobank)

century, the Tai were ready to take the reins from Angkor. The Khmer were defeated in the battle for Sukothai, and from then on, the city would be in Tai hands. Construction of buildings arranged in ways pleasing to the new masters was begun.

During Sri Indraditya's reign in the first half of the thirteenth century, Sukothai was only a small political presence, casting an apprehensive eye at the power agendas of its Tai neighbours. The turning-point in its history could be said to have arrived some decades later with the invasion of Mae Sot forces at Sukothai's own outpost of Tak. For Sri Indraditya's young son who led the troops to defeat them would soon rise, having been crowned in 1279, his greatly expanded kingdom upon his shoulders, to become a ruler of moral superiority, whose qualities of astute leadership are remembered and revered in Thailand even today.

Ramkhamhaeng's kingdom was prosperous. The famous inscription dating to 1292 states that '... this land of Sukothai is thriving. There is fish in the water and rice in the fields....' The King's subjects were encouraged to unite in a new, common identity, strengthened through the active support of Theravada Buddhism. Monks moved between Sukothai and Sri Lanka, working to spread the faith to the Tai people and their rulers. At the same time, aspects of the old Brahmanism were still accepted, and given patronage at court.

Sukothai disintegrated after Ramkhamhaeng, to be absorbed by Ayuddhya in 1431. But although the curtain was falling on the short-lived political stage of Sukothai, the art of the once great kingdom was now coming into its age of almost other-worldly originality. Although the ideal Buddha image of the Sukothai period had not yet appeared at the beginning of the age, within half a century an astonishing evolution in sculpture had taken place. And while the influence of Sinhalese Buddhism upon the Buddhism of Sukothai was undeniable, in sculpture Sukothai soared through its own visions of spirituality.

30. Sukothai-style Buddha, Bangkok National Museum. (Luca Tettoni/Photobank)

To the south, where the Chao Phya River flows silently through the flatlands, the Kingdom of Ayuddhya would have its birthplace. Both neighbouring Lopburi and Suphanburi still exercised some importance. In the field of art, Lopburi, which had once taken over the traditions of Dvaravati and known the influence of Srivijaya, at this time manifested centuries of Khmer inspiration. And although only a short distance away, Suphanburi was producing art of a different school, combining in its images of the Buddha Mon, Haripunjaya, Khmer, and Tai influences in what is known as the U Thong style. Politically, the region in the first decades of the fourteenth century lacked leadership; the time was auspicious for the right hands to seize power. Some of the factors in favour of a new political focus were the demise of Mongol China, the weakening of Sukothai, the fading glory of Angkor, and continuing conflicts in Lan Na.

Hailing perhaps from a merchant family of Phetburi, an enterprising individual named U Thong chose to be the man of the moment. Supported by his community, and married into the ruling houses of possibly both Suphanburi and Lopburi, U Thong, who would soon be known as King Ramathibodi, founded his kingdom's capital in 1351. Ayuddhya's location proved to be of strategic importance, both commercially and politically. Through the balancing of military, administrative, and socio-economic factors, the foundations for the kingdom's future strength were laid; religion would be the binding agent.

King Ramathibodi's Ayuddhya functioned differently from the traditional Tai states, its control one of organized bureaucracy rather than of personal patronage. Beginning with Ramathibodi, the Ayuddhya kings governed with a system whose origins were in Vedic India, the Laws of Manu, one of the Dharma Sastras. The atmosphere at court drew inspiration from early India and Angkor as well, with the king somewhat distanced from his subjects by his royal entourage through Brahman rituals and a Sanskrit- and Khmer-based court language.

Upon the death of King Ramathibodi in 1369, there followed decades of conflict as subsequent successors struggled to increase power, both internal and regional. During the late

31. Tonsure ceremony in Siam. (Mouhot, 1864)

fourteenth-century reign of King Borommaracha I, Ayuddhya's territories expanded to include Nakhon Sawan, Phitsanuloke, Kamphaeng Phet, and Lampang. Struggles with Angkor occupied several kings as well. Conquest of the north seemed to be the goal of the rulers who followed. Sukothai and Ayuddhya were drawn into battle as Ayuddhya intermittently attempted to establish authority there.

It was during the reign of Borommaracha II and his son that Ayuddhya rose to its position of pre-eminence on the South-East Asian panorama. Judging the foundations of Angkor to be precarious enough, the Ayuddhya king sent forth the expedition of might which would, in 1431, deliver the final blow

32. A priest in his boat. (Mouhot, 1864)

to the seat of Khmer strength. Was the Emerald Buddha there, cloistered in a chapel within the palace, then tucked into a padded box with other treasures to be carried away from the remains of Angkor to Ayuddhya, where it would reflect the grandeur and glory of that state—until it was taken elsewhere to receive the offerings of yet another town's citizens? Or had the image already been hidden amongst a priest's scanty belongings and rushed through smouldering streets to shelter in the north-west?

Fuelled perhaps by his success to the east, Borommaracha II proceeded to Sukothai, installing his son as viceroy there in 1438. King Borommatrailokanat continued to develop and strengthen the legislation system installed by his royal ancestor Ramathibodi. The efficiency of the bureaucracy would be tried during the next half century as warfare and rivalries marked the return to power of Lan Na and the emergence of the Kingdom of Lan Chang.

The art of Ayuddhya was, like its political heritage, a fusion of assorted traditions. Sukothai, Khmer, Mon, and various other Central Plains influences combined in ways which would nevertheless permit its own embodiment of expression; the Ayuddhya style would evolve over the ensuing centuries. The art of Sukothai had a profound influence, as did continued close religious ties with Sri Lanka. Later, the renewed interest in Khmer art would see crowned Buddha images. Indeed, the ornamented images of the later Ayuddhya period were far removed from the humbly garbed Buddha of earlier days.

During Ayuddhya's days of supremacy, the outlook of its rulers would change, both in how they perceived themselves and in their relations to the outside world. From its beginnings as a site on the river to a powerful empire with economic and political ties to both the east and west, Ayuddhya would remain a force to be reckoned with until the final Burmese destruction of it in 1767.

Meanwhile, behind the scenes, the day was rapidly approaching when the Emerald Buddha would make its appearance on the stage of actually recorded history.

8
Laos

Being a final glimpse of the Chronicle of the Emerald Buddha, followed by a brief recounting of some of the events which occurred in the northern Kingdoms of Lan Na and Lan Chang.

THE Third Epoch of the Chronicle opens with the accession of fifteen-year-old King Bodhisara to the throne of Luang Prabang in 1520. A devoted Buddhist, he exercised his royal prerogative by having the altars of demons cast into the river, thus ridding the city of heresies. This accomplished king later wed the daughter of the King of Muang Yuon, and their son, when sixteen, was asked to take the place of his maternal grandfather upon the throne of Chiang Mai. The boy was crowned and named Jethadhiraja. Informed of his father's death three years later, King Jethadhiraja made preparations to return to his homeland. Not knowing if he would ever return to Chiang Mai, he decided to take with him the Emerald Buddha.

The image was placed in a shrine in Lan Chang. After residing there three years, Jethadhiraja heard that a priest had been crowned king of Chiang Mai. Following an unsuccessful attempt at attacking the city, Jethadhiraja took the Emerald Buddha and settled in Vientiane. Thereafter, until his death, he led a life of devotion, and was reborn in the heaven of angels. Herewith the Chronicle closes, stating that the Emerald Buddha still resided at Vientiane, which town had become extremely prosperous.

Once again, the threads of the Chronicle and what is today accepted as historical fact cross at certain points. At the time of the discovery of the Emerald Buddha, Chiang Rai was under the rule of Sam Fang Kaen. This king spent his reign strengthening Lan Na which, after a period of difficulty upon the death of King Mengrai in 1317, had been working to regain political and cultural vitality. Sam Fang Kaen's grandfather, King Ku Na, had injected the religious and cultural life of the north with new intellectual force through the establishment of a Sinhalese Order. How pleased his grandson must have been to find such an auspicious treasure within his own kingdom.

Wanting to enshrine the Emerald Buddha in Chiang Mai, King Sam Fang Kaen commanded the image be brought to him there. But once again fate would have it differently. An elephant was sent to collect the Emerald Buddha but when the procession arrived at a junction where the road led to Lampang, the elephant turned and steadfastly moved on to that town. Two more attempts were made, and each time the elephant headed towards Lampang. Such an omen could not be ignored. The Emerald Buddha was duly installed in Lampang, where it remained until 1468.

By then, a new ruler was on the throne of Lan Na, a king who would surpass his father in action. The reign of King Tilokaracha was marked by defensive and offensive warfare (especially with the forces of Ayuddhya's King Borommatrailokanat), espionage, and a determination to make Lan Na a formidable power. On the spiritual level, King Tilokaracha ensured that his kingdom would see Buddhism reign supreme. In AD 1468, the Emerald Buddha was at last taken to Chiang Mai where it was installed in the great chedi there. Religious contacts between Lan Na and Sri Lanka were strengthened so that the Buddhism of King Tilokaracha's realm should be as pure as that of the Sinhalese version. A cutting was brought from the sacred bodhi tree at Anuradhapura and planted at a monastery in Chiang Mai. During his reign, a small-scale copy of the revered Mahabodhi temple at Bodhgaya was built to celebrate two thousand years of Buddhism.

Despite skirmishes with Ayuddhya and involvement with the Shan States, the Emerald Buddha watched over a predominantly stable Lan Na. During King Muang Kaeo's time in the early part of the sixteenth century, much emphasis was placed on supporting the intellectual and cultural dominance of the Sinhalese order of Buddhism. The glory days of Lan Na were coming to an end, however. Political rivalries and violence, both from at home and further afield, wrought great damage to the institutional structure of the kingdom. It was at this point that the young prince of Luang Prabang mentioned in the Chronicle was invited to rule from Chiang Mai.

In the middle of the sixteenth century, Luang Prabang, capital of the Kingdom of Lan Chang, was confident of its wealth and independence. Its strength in the region, however, was only a relatively recent development. The preceding centuries had seen the central Mekhong valley influenced by Dvaravati, Angkor, and Sukothai. In 1353, only two years after the founding of Ayuddhya, a young prince, born in exile in Angkor, had fought his way back to his paternal home to unite some of the Laos states. From upon his throne at Lan Chang, Fa Ngoun expanded the borders of his kingdom until it was one of the largest of the time. Notwithstanding his interest in military matters, Fa Ngoun practised Theravada Buddhism, which he had been advised to do by his father-in-law, the Khmer King Jayavarman Pararmesvara. Together with his wife, King Fa Ngoun encouraged his subjects' conversion as well. A religious mission was sent to Cambodia from which Buddha images and holy scriptures were brought back. Amongst these was the Pra Bang, said to have been a gift to Angkor from a king of Sri Lanka. This image of the Buddha was given a place of honour in the capital. Indeed, later the city itself would be named after it.

Sam Saen Tai took over from his deposed father, Fa Ngoun, in 1373 and saw to it that during his reign the kingdom knew peace and a flourishing of Buddhist institutions. With its ruler married to a princess of Ayuddhya, Lan Chang enjoyed good relations with that kingdom, gaining commercial prosperity as well as knowledge of Ayuddhya's methods of state organization. In its art and architectural forms, too, Lan Chang borrowed much from its Siamese neighbours. Alliances with the ruling houses of Chiang Hung and Lan Na further strengthened Lan Chang's position in the region. Although subsequent decades were marked by internal power struggles followed by serious altercations with a now hostile Annam, by the time of T'ene Kham, probably

33. A Buddha within Ho Phra Kheo, Vientiane. (Steve Van Beek)

around 1479, the land was once again beginning a period of peace.

During the years up to the reign of King Phothisarat beginning in 1520, Lan Chang developed close ties with Annam and the cities of the Menam valley. At the same time, a Lao identity,

34. Panel in That Luang, Vientiane. (Steve Van Beek)

separate and apart from that of its neighbours, however goodwilled they might be, was encouraged. This spirit of independence and self-confidence was at its height during the time of King Phothisarat who, for reasons of commercial and administrative convenience, moved his capital from Luang Prabang to the more centrally located Vientiane. As mentioned in the Chronicle, this was probably the ruler who ordered destroyed the buildings dedicated to traditional animistic and even Brahmanical worship in a not completely successful attempt to outlaw observances other than those of Theravada Buddhism.

King Phothisarat cultivated diplomatic relations with Yunnan, Champa, and Annam, and matrimonial alliances with Lan Na

35. That Luang, restored in the eighteenth or nineteenth century, Vientiane. (Steve Van Beek)

and possibly even Ayuddhya and Cambodia during this era of Lao supremacy. But old patterns would repeat themselves in the tapestry of time. Beginning with the reign of his son Setthathirat, the fortunes of Lan Chang changed with the rapidly altering shape of the surrounding world. Following Setthathirat's return to Vientiane with the Emerald Buddha after his accession to the throne of Chiang Mai, touched upon in the Chronicle but not ending as peaceably as therein stated at the time, the region was beset by interlinked happenings of a tumultuous nature.

Lan Na was soon the site of civil unrest. Prince Mekuti of the Shan state of Nai, a descendant of King Mengrai, came to the Chiang Mai throne in 1551, yet within a few years was but a vassal of the Burmese who were now a major power. By 1564, the great Kingdom of Lan Na was little more than a military base, used by the forces under King Bayinnaung to mount Burmese operations against both Ayuddhya and Lan Chang. Ayuddhya was a natural attraction to the Burmese, its access to the sea promising control of maritime trade, and would over the centuries be a prize often contested. After a long siege, in 1569, a vassal king from Phitsanuloke was installed by the Burmese upon the throne of Ayuddhya.

Now only Lan Chang remained as a power. In 1564, King Setthathirat had transferred the Emerald Buddha from Luang Prabang to Vientiane, where it would come to rest in the newly built Vat Pbrapha Kheo. During the next decade, in which the kingdom was able to escape Burmese control, several Buddhist shrines were erected by King Setthathirat, landmarks such as the That Luang stupa in Vientiane, while others were renovated. But 1570 brought the beginning of the end for Lan Chang. In that year, the King disappeared during a military campaign, and during the scuffle over the succession the Burmese were able to take power in 1574. Theirs would be a rule unlike that of the fallen kingdoms, but the challenge would be met by a spirit of survival. Its gaze enigmatic, the Emerald Buddha would watch the events of the following two centuries.

9
To Bangkok

Being a sighting of the Emerald Buddha during its brief sojourn in Thonburi, as well as a glimpse of the voyage across the waters of the Chao Phya to its final resting place in the Grand Palace of Bangkok.

FOR two hundred and fourteen years the Emerald Buddha remained in Vientiane. By 1591, Laos had won independence under King Nokeo Koumane. Dynastic struggles followed and it was not until the firm and lengthy reign of Souligna Vongsa, beginning in 1637, that the kingdom enjoyed internal peace as well as predominantly friendly foreign relations. With Souligna Vongsa's death in 1694, rivalries of succession caused the split of the kingdom into two hostile states with capitals at Luang Prabang and Vientiane.

Delicate dealings with Burma and Annam allowed Vientiane to outweigh Luang Prabang for some time, but by 1774 the tables had turned. Phraya Taksin, then a governor of Tak, had been working successfully to drive the Burmese out of Ayuddhya, which they had again captured in 1767. When seven years later Luang Prabang allied itself with Taksin, Vientiane was doomed. In 1778, under the command of Chaophraya Chakri, a nobleman of Ayuddhya, Siamese forces captured the city. Taking with them both the Emerald Buddha and the Pra Bang (the latter would be returned to Laos in the following century), the victorious troops returned to Siam. With the erstwhile kingdoms and principalities in which the Emerald Buddha had once resided now incorporated into the newly formed kingdom, Siamese sentiment must have been that the venerated image was coming home. The destination, however, was not Ayuddhya but a port on the western bank of the Chao Phya, Thonburi. There, the erstwhile General Taksin, with the support of the Chinese trading community of the region, had been able to establish the base which would be his capital during the action-packed years of

36. Wat Arun. (Mouhot, 1864)

his power. The Emerald Buddha was installed at Wat Arun.

Crowned in 1768, King Taksin accomplished much militarily, and in his day the Kingdom of Siam grew to a size never before seen. Nakhon Si Thammarat, Phitsanuloke, Fang, and most importantly, Lan Na, were among the rival states drawn into the realm. But by 1779, Taksin's mental stability was in question as he demanded of the Sangha that he be worshipped as a deity.

37. A Siamese prince and attendant. (Thomson, 1875)

38. View of the port and docks of Bangkok. (Mouhot, 1864)

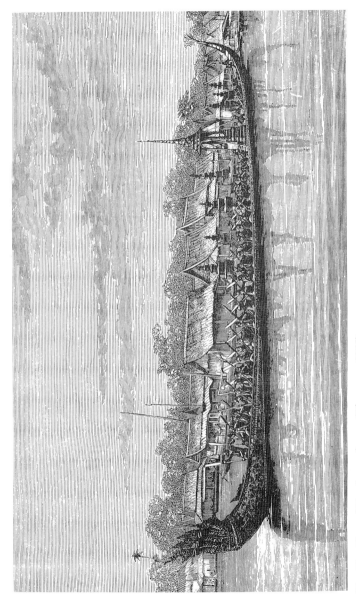

39. The King of Siam's state barge. (Thomson, 1875)

Two years later, Taksin sent forces under Chaophraya Chakri to Cambodia to settle a power struggle there. During this mission the disapproval which had been building up in the Thonburi court over the King's behaviour erupted, and he was overthrown. Upon his return, Chaophraya Chakri was invited to take his place upon the throne at Thonburi.

Perhaps to erase some memories of his predecessor, King Rama I, first ruler of Thailand's present Chakri Dynasty, moved his capital across the water to a new city. Bangkok, or Krunthepmahanakornratanakosin as the King named it (the latter 'Ratanakosin' meaning 'Repository of the Gem Image'), was constructed partly with bricks shipped down the river from the ruins of Ayuddhya, a fusion of memory and modernity. In 1784, an abode deemed suitable to honour the image of the Buddha, the palladium of the Thai nation, was ready. On 22 March of that year, the Emerald Buddha was taken across the river, perhaps in a grand procession of royal barges little different from that which occasionally takes place upon the Chao Phya today. Once in the compound of the Grand Palace it was carefully ensconced in the Chapel Royal, whose design is said to have been inspired by the seventh-century Indian temple at Mamallapuram.

In the centuries following its installation, the Emerald Buddha has watched over the Thai nation and, in turn, been revered by its folk. Rama I re-established and restored discipline to the Sangha, and under his patronage the Buddhism of Siam flourished. Indeed, Rama I's actions, religious and legislative, were deeply rooted in Theravada Buddhist and Siamese traditions, although with spirit and innovations mindful of both past and future.

Successive kings would inhabit a world whose perimeters were ever widening. Where once the skills of statesmanship had been of value in dealing with neighbouring Tai states or kingdoms in the South-East Asian region, the kings of the Chakri Dynasty would learn to secure peace for their people in a greatly expanded political arena. In turn, each king graced his subjects with a gift unique to his own reign. From Siam to Thailand, today the land of the Mon, Khmer, Tai, and other

40. Wat Phra Keo. (Vincent, 1874)

ethnic groups is a nation proud of its heritage, conscious of its role in the continuously evolving tapestry of tomorrow's world. The Emerald Buddha is in the design, held close to the heart of the Thai people.

Nagasena's invocation on the full moon night of the Emerald Buddha's consecration has been fulfilled; Buddhism has indeed been given 'brilliant importance in five lands'.

Epilogue

IN an era of buildings reaching boldly into the sky, their façades sleek and hurried, the ornate walls of the Chapel Royal are a reminder of other days: an age when time was measured in its most natural rhythm, the darkness and the light, the seasons. The infinite detail here, the countless millimetres of glittering adornment, are the creation of hands seemingly fearless of the passing of a lifetime.

Inside the *ubosoth*, away from the sunbaked heat of the morning, myriad figures hover in scenes from the *Ramakien*. The murals tell of the Buddha, and his search for Truth; the universe. The walls vibrate, too, with the voices of those who have come, each for his own reason, to be in the presence of the image. A group of monks from a distant land, the hems of their pale grey robes just sweeping the mosaic, suddenly move in unison to touch their heads to the floor. Near by, a husband and wife unpack offerings they have brought with them in a basket, and the pungent scents of egg and pickled vegetables are freed as the food is placed into one of the receptacles positioned along the railing separating the visitors and the altar of gold. Drawn to the focal point, a woman of the West sits in reflective silence.

Towering high above the constantly changing sea of bodies is the Emerald Buddha. The smooth solidity of its form is of a green both warm and cool, and flecked as if with galaxies of dark stars. Although covered with the intricate golden finery of the Hot Season garments, the broad shoulders crowned by glittering leaves pointing upwards and away, the body of the image radiates strength. Whose inspired hands carved these lines, felt the essence of eternal creation in the raw gemstone? In what tongue did he speak? Will we ever know with certainty? From a certain angle the bejewelled eyes, darker at the pupils, seem to look into the viewer, and much further, an expression beyond emotion; patience, infinite. One could get lost in this vision or, perhaps, found. The answer to the mystery of the Emerald Buddha is contained within it.

Select Bibliography

Anonymous, *The Chronicle of the Emerald Buddha*, translated from the Thai by Camille Notton, Bangkok, 1932.

Basham, A. L. (ed.), *A Cultural History of India*, New Delhi, Oxford University Press, 1975.

———, *The Wonder That Was India*, 3rd edn., New Delhi, Rupa and Co., 1992.

Bibby, Geoffrey, *Looking for Dilmun*, London, Pelican Books, 1972.

Bock, Carl, *Temples and Elephants: Travels in Siam in 1881–1882*, London, Sampson, Low, Marston, Searle, & Rivington, 1884.

Boisselier, Jean, *The Heritage of Thai Sculpture*, Bangkok, Asia Books Co. Ltd., 1987.

Coedès, G., *The Making of South East Asia*, Berkeley, University of California Press, 1967.

Craven, Roy C., *A Concise History of Indian Art*, New York, Praeger Publishers, 1976.

De Silva, K. M., *A History of Sri Lanka*, Berkeley, University of California Press, 1981.

Fickle, Dorothy H., *Images of the Buddha in Thailand*, Singapore, Oxford University Press, 1989.

Hall, D. G. E., *A History of South-East Asia*, 2nd edn., London, Macmillan and Co. Ltd., 1964.

Kelly, R. Talbot, *Burma*, 2nd edn., London, A & C Black, 1933.

Ludowyk, E. F. C., *The Story of Ceylon*, London, Faber and Faber, 1962.

MacDonald, Malcolm, *Angkor and the Khmers*, London, Jonathan Cape, 1958.

Mouhot, M. Henri, *Travels in the Central Parts of Indo-China (Siam), Cambodia, and Laos . . .* , London, John Murray, 1864.

Rawson, Philip, *The Art of Southeast Asia*, London, Thames and Hudson, 1967.

Singhal, D. P., *India and World Civilization*, Calcutta, Rupa and Co., 1972.

Sirisena, W. M., *Sri Lanka and South-East Asia*, Leiden, E. J. Brill, 1978.

Smith, Vincent A., *The Oxford History of India,* 4th edn., New Delhi, Oxford University Press, 1988.

Thomson, J., *The Straits of Malacca, Indo-China and China...,* London, Sampson, Low, Marston, Low, & Searle, 1875.

Thongsib Suphamard, *Phra Kaeo Morakot* (in Thai), Bangkok, Silapakorn University.

Vincent, Frank, *The Land of the White Elephant...,* New York, Harper & Brothers, 1874.

Wyatt, David, *Thailand: A Short History*, New Haven, Yale University Press, 1984.

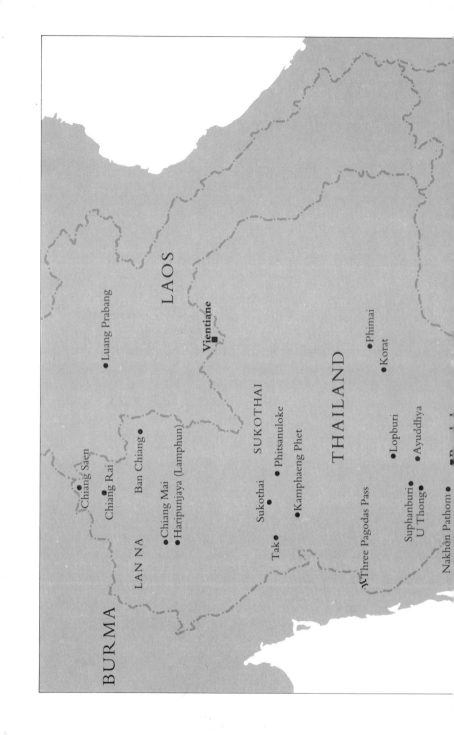